Life as a Geological Force

The Commonwealth Fund Book Program
gratefully acknowledges the assistance of
The Rockefeller University
in the administration of the Program

A volume of
THE COMMONWEALTH FUND
BOOK PROGRAM

Under the editorship of Lewis Thomas, M.D.

PETER WESTBROEK

Life as a Geological Force

DYNAMICS OF THE EARTH

W · W · NORTON & COMPANY

NEW YORK · LONDON

FIRST EDITION

THE TEXT OF THIS BOOK *is composed in Baskerville, with the display set in Typositor Cactus. Composition and manufacturing by The Maple Vail Book Manufacturing Group. Drawings by Cees van Nieuwburg. Book design by Marjorie J. Flock.*

Library of Congress Cataloging-in-Publication Data
Westbroek, P. (Pieter), 1937–
Life as a geological force: dynamics of the earth / by Peter Westbroek.
p. cm.—(The Commonwealth Fund Book Program)
1. Geobiology. I. Title. II. Series: Commonwealth Fund Book Program (Series)
QH343.4.W47 1991
551—dc20 90-34862

ISBN 0-393-02932-8

W.W. Norton & Company, Inc., 500 Fifth Avenue, New York, N.Y. 10110
W. W. Norton & Company, Ltd., 10 Coptic Street, London WC1A 1PU

1 2 3 4 5 6 7 8 9 0

TO OUR DAUGHTERS

TO MY MOTHER

CONTENTS

Forcword 11
Preface 13

PROLOGUE

1. Peculiar Planet, Familiar Ground 19

I. Geology of the Planet Earth

2. Shifting Plates 47
3. Stability and Change 69
4. The Stylized World of Robert Garrels 89

II. Life, the Missing Link

5. Life and the Cycling of Rocks 111
6. The Power of the Small 136

7. A Rock Evolves 147
8. Humble Foundations 166

III. Life at the Planetary Scale

9. Biospheres 183
10. Gaia and the Frankenstein Monsters 204

EPILOGUE

11. Global Change and the Melding of Sciences 219

Suggestions for Further Reading 229
Index 233

FOREWORD

"THE ADVANCE OF SCIENCE, during just the past half century, has been an amazing achievement and a totally good thing for the human species, carrying all manner of promise for the years ahead." Assertions like this are made every day within the scientific community, and are taken for granted all around.

But still there hangs in the public mind a set of misgivings about the whole enterprise: Where are these people taking us? What new technologies will they next be introducing to change the way we live? (And most troubling of all) Are they explaining away all the old mysteries, flattening out the beauty of nature, displacing strangeness by their reductionism?

This book by Professor Westbroek provides, in my view,

an exhilarating response. Because of the new geology and biology, the life of the earth has become a far stranger and more splendid puzzle than ever before. We are only beginning to comprehend how little we really know about it; we confront deeper mysteries at every bend as we begin to learn about interactions among living things on the planet and their profound effects on the very structure of earth. To be sure, scientists have taken us a certain distance, but everything is still to be learned. At the end—if there is an end—is the lively possibility that the life of the earth is all of a piece, an almost impossibly intricate organism in itself.

The Advisory Committee for The Commonwealth Fund Book Program, which recommended the sponsorship of this volume in its series of books written by working scientists about their own fields, consists of the following members: Alexander G. Bearn (deputy director); Lynn Margulis, Ph.D.; Maclyn McCarty, M.D.; Lady [Jean] Medawar; Berton Roueché; Frederick Seitz, Ph.D.; and Otto Westphal, M.D. The publisher is represented by Edwin Barber. Margaret E. Mahoney, president of The Commonwealth Fund, has actively supported the work of the Advisory Committee at every turn.

Lewis Thomas, M.D., Director
The Commonwealth Fund Book Program

PREFACE

THE PROBLEM OF environmental degradation is nearly as old as humanity, but never before has public concern been as deep and widespread as it is today. For the first time, general anxiety transcends personal and national boundaries. The well-being of the earth as we know it is at stake. The media reverberate with ominous premonitions about the future habitability of the globe, and the problem is high on the agenda of the top world leaders. How long will we last, and what can we do? Is the environment as ill as it seems, or are we entangled in a collective delusion?

Science and technology are seen as the root of the problem, yet, ironically, they are also expected to rescue us. The environmental fever is sparking worldwide research efforts. We hear huge programs described as "unprecedented in [their] comprehensive interdisciplinary scope, to address the functioning of the Earth system and to understand how this system is changing." The information gathered will "help provide the world's decision makers with input necessary to wisely manage the global environment."*

Unfortunately, science is poorly prepared for this venture. The level of integration required to understand the earth system is far beyond our present reach. The sciences have been divorced for a long time and their reunion is a long and painful process, comparable to the integration of different cultures and races. Therefore, noble initiatives for the study of global change are unlikely to reach their lofty goals in the short term. Yet, they mark a change of heart and a new beginning.

This book attempts to probe the prospects for a comprehensive science of the earth and concentrates in particular on the melding of two fields of vital importance: geology and biology. It exposes deep rifts that divide the ideas, traditions, and work styles of these sciences and explores the enormous potential of their integration. The long-neglected idea of life as a major geological force, with a profound influence on the dynamics and history of the earth, is reemerging as the critical concept by which to assess the integration between the earth and life sciences. At it turns out, the impact of life is widely acknowledged (although poorly understood) as long as the present-day earth is studied at the local or regional scale. When we turn our attention to the geological past, our recognition of the active involvement of life fades away. The confusion is total as

* International Geosphere-Biosphere Programme (IGBP), *A Study of Global Change* (Stockholm: The Royal Swedish Academy of Sciences, n.d.).

soon as we take an astronomical view of our planet, and consider the earth in the four dimensions of space and geological time. With such a confused view of the earth, how can the world's decision makers ever manage the planet?

A scientific discipline resembles the neuronal network of the brain, a delicate and immensely complex maze of connections constantly building new ramifications and enhancing its organization. The merging of two such networks is an absorbing and fascinating spectacle that no book can describe in a straightforward and consistently focused way. I chose to highlight a few subjects with which I am to some degree involved and that I consider to be especially significant. The chapters that follow are self-contained; not much harm is done when the order is changed. Underlying the book is a subtle organization, however—a line of thought best discerned when the reader follows the given succession of chapters. In the personal view that emerges, I do not pretend to reveal more than a few outstanding aspects of a huge scientific development.

I hope to reach a broad readership: policymakers, teachers, students, and experienced professionals who wish to speculate about the nature of the planet on which we live and to evaluate critically the state of our knowledge. I hope this book will help those who are involved in building a more comprehensive science of the earth.

This book would have never been written without the support of Lewis Thomas and The Commonwealth Fund Book Program, which have been both generous and patient. I am especially grateful to Lynn Margulis, who convinced me to write the book, and who time and again helped restore my self-confidence. My contacts with Alexander Bearn, Helene Friedman, Nina Bouis, and Evelyn Conrad are also highly appreciated.

An unusual amount of editing was required to bring this hybrid book into its present format. I am very grateful to Jeremy Young and Mirjam Westbroek, and particularly to Jackie Wilson and Cheryl Simon Silver, for their skilled and extensive editorial help. Philip Sandberg and Matthew Collins spent many hours correcting the manuscript.

Thomas D. Brock and Frits Böttcher helped me over my early reservations. Gerrit Jan de Bruyn, Simon Conway Morris, Gordon Curry, Hans van Gemerden, Ricardo Guerrero, Hans Kamerling, Michael LaBarbara, Lucas Stal, Jerome Ravetz, Sieband P. Thallingi, Fried Vogels, and Jaap Westbroek are gratefully acknowledged for their expert advice. It was a particular pleasure to collaborate with Cees van Nieuwburg, the artist who made the fine illustrations.

I would like to thank my colleagues and students in the Geobiochemistry Research Group in Leiden. Generous support by Alan McHenry and his board of directors at the Richard A. Lounsbery Foundation of New York City greatly stimulated our research despite the time I had to devote to the book.

Throughout the work I was inspired by that moment when, as an eight-year-old boy, I walked through the dunes with my father. "Look around you," he said. "All you can see is really there. Yet, behind each thing lies a mystery, hidden from view. If you try hard, you may get a glimpse of that world, but never will you be able to see it all."

Life as a Geological Force

1

PROLOGUE

PECULIAR PLANET, FAMILIAR GROUND

More than thirty years ago, when I was in my teens, our next-door neighbor, a lady of wealth and dazzling beauty, gave me an unexpected glimpse into her intimate feelings, and sighed: "Peter, you have no idea how lucky you are to be young. I get older and older and may not be there when man sets foot on the moon. All I want from life is to witness that climax of history, and then I won't mind if the Lord takes me away."

As it happens, she's still here, slightly older and surely wiser. Her words, however extravagant, expressed the general mood of the fifties—at least in Western Europe. The horrors of World War II were fresh on everyone's mind, but a new and better world beckoned. Inexhaustible natural resources invited exploitation. Cars and washing machines came within reach of the people. Factories were constructed, cities expanded. Plastics, television, and other intriguing new products conquered the marketplace.

Space travel held the greatest promise. If we humans could move out into space, surely we would overcome the trivial problems remaining on earth. Humankind would fulfill its destiny: to dominate the natural world. Maybe we would conquer the universe.

Among the finest accomplishments of the voyages to the moon were photographs of our own earth. Ironically, those pictures, which symbolized human triumph over nature, also marked the end of the general sense of promise and progress. The sixties were winding down, and in those snapshots we now read a message of global vulnerability and human impotence. What were we doing to our planet? In the years that followed we grew increasingly aware of pollution, acid rain, desertification, the rise of greenhouse gasses, and the disintegration of the ozone shield. And we began to wonder: Is this the fruit of our naive endeavor?

As the twentieth century draws to a close, it is clear that the natural order is being transformed and subjected to the economic, social, and military needs of a divided humanity. The pictures of the earth frame our problems in a global context. Is the planet ill? Is there a future for us? How will the planet respond to our activities?

We cannot yet answer these questions. If we are to predict the changes in store for our planet we need to know how it works. However enlightening they are, pictures from space are misleading. They show us the earth as a three-dimensional object but fail to reveal the fourth dimension: time. A comprehensive, four-dimensional understanding would account for the evolution of our planet throughout the 4.6 billion years of its existence. This prospect represents a major challenge for science. We must unravel the complex interactions between physical, chemical, biological, and cultural forces that have engendered this highly anomalous object in space. At present, however, not only humanity but science too is divided: today science is split into disciplines individually incapable of explaining how the earth became, and continues to function, as a single integrated system.

The foundations for synthesis and cooperation were laid by earlier generations of scientists. In particular, the great Russian scientist Vladimir Vernadsky (1863–1945), without the help of glamorous pictures from space, developed an embracing perspective of a dynamic earth. Vernadsky was largely ignored in his day, and only now are we beginning to recognize the significance of his work.

With mounting evidence that the lots of human destiny are cast together, the classical distinctions between scientific disciplines are beginning to blur. Until recently, for instance, geologists overwhelmingly interpreted the geological record in purely physical terms. The folding and faulting of rocks, their melting and solidification, the sorting of particles in running waters—those were the kinds of age-old processes that geologists could best recognize. Then they began to consider chemical aspects, and eventually new disciplines such as geophysics and geochemistry emerged, bringing powerful new approaches and technologies. Now, geologists are beginning to concede that their ancient artifacts embody signs that life itself has strongly affected the very conditions that allowed it to take hold on our planet and persist for nearly 4 billion years.

Knowledge of the biological contribution, of course, still

lags behind. We begin to appreciate the effects of life in the present environment, but as yet little of our information can stand the acid test of geological theory. Life must have been very important in the past, but we are poorly informed about *how* its effects can be recognized in the geological record. A comprehensive geological theory of the human effects is even further away.

The approach of this book reflects both the fragmented state of science today and the social and intellectual currents that converge to bring disparate sciences together. Through a series of examples I will explore the dynamics of our planet as explained by classical geology, as well as by the less known but undeniable fact that living forms help shape the earth's geology. Today, the integration of geology and biology is fashionable among science managers, but in reality that integration is most difficult to achieve. All we have learned so far is how ignorant we are about the most fundamental aspects of the dynamics of our planet. I shall concentrate on earth dynamics, and only allude to the geological impact of culture, this new, Promethean power that appears to be propelling us willy-nilly into an unknown, precarious future.

A Stroll through the Landscape

The landscape of Nieuwkoop, a village in my native Holland, provides a ready example of ways in which natural and cultural forces combine to affect and alter geological history. At first it may seem as if biology came into play only after favorable conditions had been set by physical and chemical factors, such as the formation of a lagoon protected from marine incursions by a coastal barrier and the establishment of a suitable climate with moderate temperatures and adequate water supplies. But life strongly influences the formation of lagoons in this kind of region, and living systems profoundly affect even the water cycle and climate. Finally, 800 years of exploitation and environmental management have given this region its present highly artificial character.

The name Holland, derived from *Houtland* ("Woodland") now designates two major provinces, North and South Holland, in the west of the Netherlands. The landscape is familiar from paintings by great Dutch masters such as Rembrandt, van Goyen, and Ruysdael. Their subjects—green pastures, ditches, dikes, and towns—rest largely on polders: stretches of land reclaimed from the waters that used to ravage the area.

Near Nieuwkoop, the course of the Meije, a little stream winding through the meadows, is about the only feature that has not been altered dramatically over the centuries. Along the Meije, one can ride a bicycle on the road atop the bordering dike. On a cool summer day a steady breeze blows from the northwest and carries gray, bulky clouds across the sky. This is a wet and windy country, and one is wise to bring a raincoat. The Meije is on the left, and on the right, at the foot of the dike, is a long string of farmhouses, many of which are beautifully preserved specimens from the seventeenth century. Several of them have reed roofs, and they are surrounded by pleasant orchards, shrubs, and vegetable gardens. Beyond the farmhouses, narrow canals separate long, green meadows, forming parallel rows that stretch away from the road. The meadows are often about a hundred yards wide and more than half a mile long. At their far end, away from the road, rows of trees pleasantly interrupt the otherwise monotonous landscape. Beyond are more parallel strips of meadows, and at the horizon one can discern the towns of Woerden and Bodegraven.

To the left, on the other side of the Meije, a similar pattern of elongated meadows and canals trends away from the river (fig. 1.1). The landscape here is more loosely organized. The bushes at the end of the meadows are more haphazard then the trees on the other side, and the canals are not so straight. Reeds grow along the water course, together with patches of pollard willows and a wealth of water flowers. Cows stare at you peaceably as they ruminate.

At the café de Halve Maan I used to get off my bicycle, drink a cup of excellent coffee with Jaap Schutter, the proprietor, and rent a rowboat. I crossed the Meije and rowed along one of the perpendicular canals towards the bushes. The meadows are soaked, and nearly level with the water in the canal. They give way to fields of reeds, mosses, and flowering plants: real boglands. The area is cut into a system of regular patches by a rectangular network of canals, ditches, and stands of alder, birch, and willow. This is a paradise for botanists and ornithologists.

The canal now widens and flows into one of the many lakes in the area. This one is a mile or so across and about 12 feet deep. Reeds and clusters of trees line the shore. Across the lake is a complex labyrinth of narrow ridges of land alternating with waters up to a hundred yards wide. Again, the land ridges are covered with reeds and trees. Then comes the sprawling village of Nieuwkoop, a 700-year-old settlement, now a local water-sports center. Originally, the village formed a ribbon along the dike that cuts off the bog area, and many of the seventeenth-century buildings are still intact.

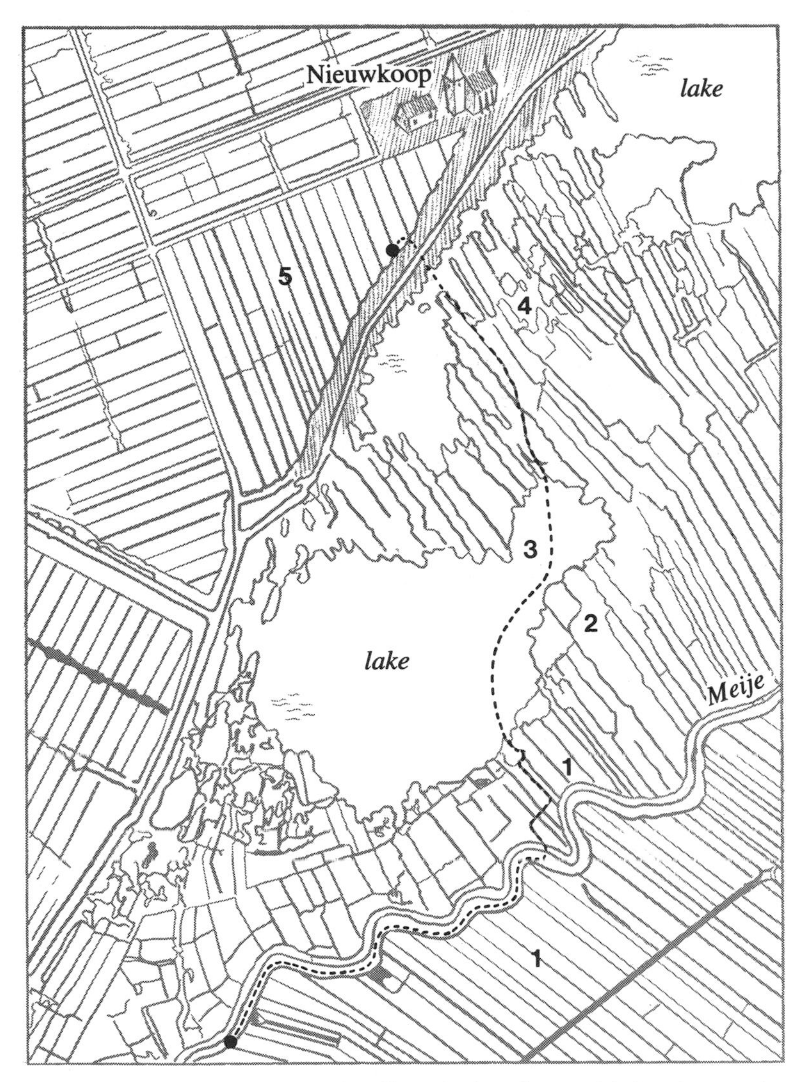

1.1 Map of the surroundings of Meije and Nieuwkoop, with five characteristic landscapes: (1) farmlands exploited since the Middle Ages; (2) reedlands exploited since the nineteenth century; (3) a lake; (4) peat exploitations (seventeenth century, mainly); (5) the nineteenth-century polder at Nieuwkoop.

Beyond the dike lies a huge polder that, until a century ago, was a lake. It is at least 12 feet below the surface of the waters we have just crossed. Blocks of rectangular meadows are separated by meticulously arranged sets of perpendicular roads, dikes, rows of trees, and large farmhouses. The scale is larger and the planning more efficient and modern than in the historic panorama with which we started our trip.

This little excursion has carried us across five types of landscape characteristic of the western region of the Netherlands: farmlands exploited since the Middle Ages; reedlands that originated in the nineteenth century; a lake; peat exploitations (seventeenth-century, mainly); and the nineteenth-century polder at Nieuwkoop. It is a carefully designed system of multilevel waterways, polders, and dikes—the result of a struggle of centuries between humanity and the elements. There is nothing purely natural. If left unattended the whole area would soon be flooded by the sea.

An interesting paradox underlies these terrains. A thousand years ago this was wilderness, a virtually impenetrable and uninhabitable region between the sandy hills to the east and the low sand dunes along the coast. In the seventeenth century, however, this very area provided the economic basis for a mighty empire: Holland in the Golden Age.

Natural Causes

To understand the development of this region one must go back more than 10,000 years, when the last ice age ended and the present Holocene period began. Figure 1.2 shows the area in the larger geographical context. The major part of the Dutch territory can be viewed as a river delta merging into a sea whose floor is dropping. Over more than a million

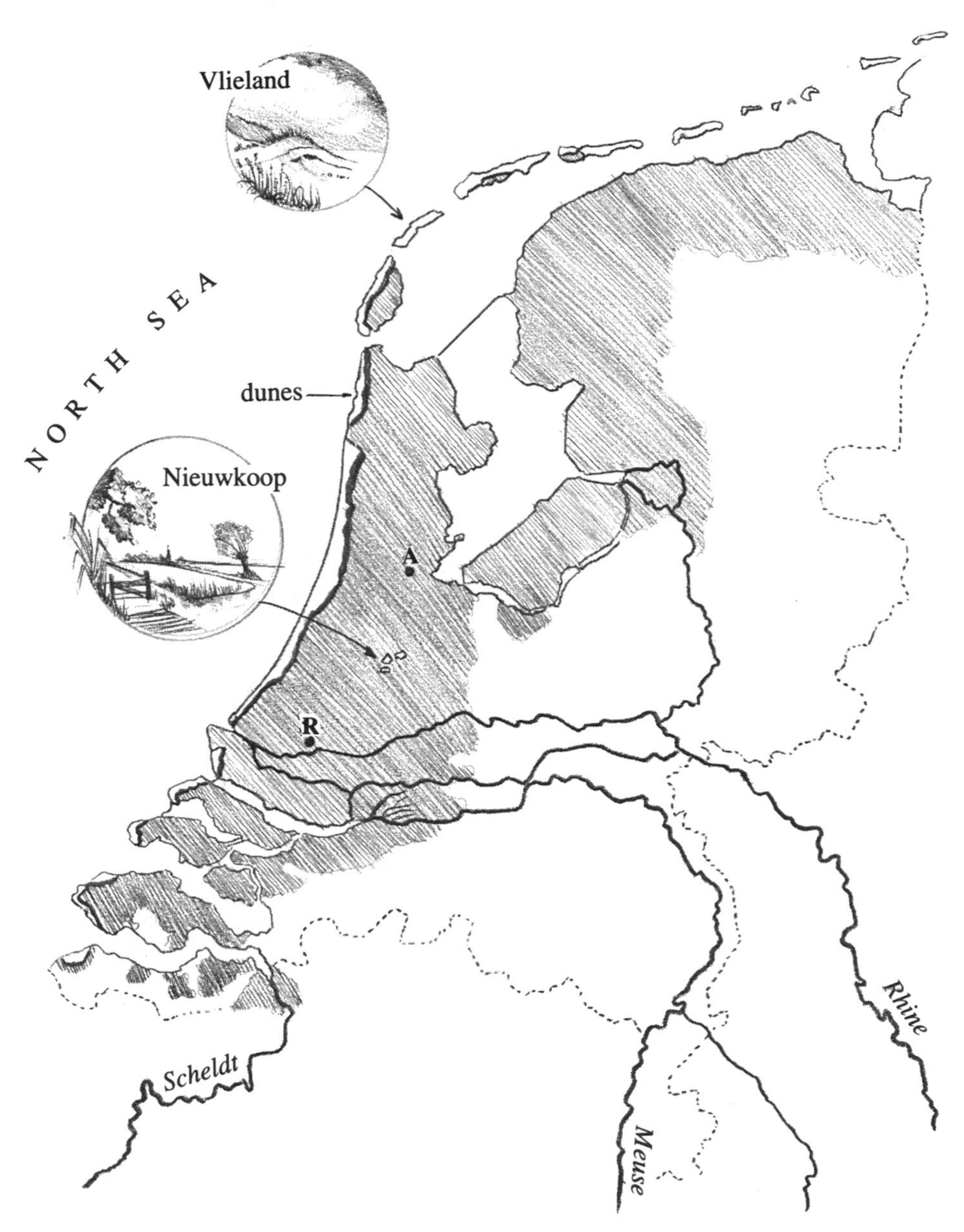

1.2 Map of the Netherlands, including the delta of the rivers Rhine, Meuse, and Scheldt. The shaded area is lower than mean sea level plus 1 meter, and would be flooded if left unprotected. Note the position of the dunes along the coast, the mud flats between the string of islands in the north and the mainland, and Nieuwkoop and Vlieland (discussed in chapters 1 and 5, respectively).

years, the rivers that flow down into this delta, particularly the Rhine and Meuse, filled it with sands and clays eroded from the weathering mountains upstream.

During the last glacial period, ice covered large continental areas around the Arctic; its southern boundary ran across northern Germany and southern Denmark, and into the North Sea. So much water was tied up on land that the level of the sea was some 300 feet lower than today. Much of the basin now filled with the North Sea was dry, and in the Netherlands a polar desert or a tundra-like regime prevailed. Most of the land was covered with sands brought down by the rivers and tossed around by the wind.

When temperatures moderated at the beginning of the Holocene, the ice caps started to melt. The sea level rose, and about 7000 years ago reached the present Dutch coastline. The sea went even further inland and then was pushed back again by the steadily accumulating sediment. The rivers brought down huge masses of clays that were swept into the sea and accumulated in a thick blanket along the coast—a huge mud flat that widened over time, edging toward and eventually joining the land.

About 3000 years ago, low sand dunes started to develop along the western and northern coast of the Netherlands, protecting the inland region from marine incursions. At this stage a zone behind the dunes, 10 to 30 miles wide, was transformed from mud flats into a huge marshland where large masses of peat could accumulate. The wilderness was born. This sequence of events is easily reconstructed from the sediment distribution shown in cross section in figure 1.3.

Peat is a wet accumulation of the remains of plants and sphagnum moss formed when the rate of plant formation exceeds that of its decomposition. The major causes of the decomposition of plant debris are bacteria and fungi. As these consumers dissolve organic plant material, carbon dioxide and water are released. Oxygen is used up in this process. This gas is very poorly soluble in water, however, and microbes attacking plant debris in a moist environment rapidly remove it. As a result, the water-soaked sediments become oxygen-depleted, or anaerobic, which slows down or even halts the activity of the consumers. Bacteria that can live under anaerobic conditions may catalyze some further breakdown of organic matter, but at a much reduced pace. Moreover, they release organic acids and carbon dioxide in the process. The ensuing acidity may slow biological activity, thereby further delaying decomposition.

Production and, especially, destruction of plant material are strongly stimulated by high temperatures. In the tropics, plant destruction tends to outstrip formation; in polar regions, the rate of plant production is so low that

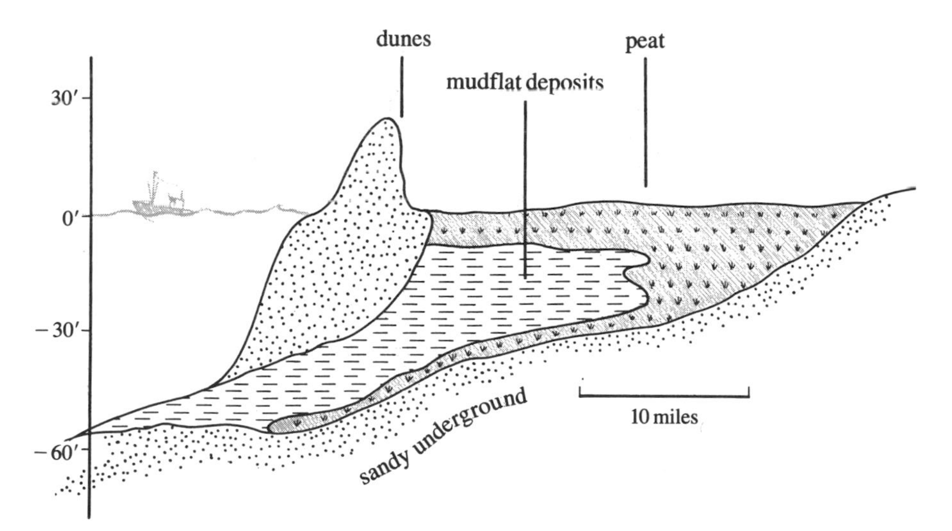

1.3 Cross section through young sediments in the western part of the Netherlands, showing extensive peat development behind the coastal dunes.

accumulation cannot take place on a large scale. Consequently, the optimum conditions for peat formation are in the temperate zones. Throughout the Holocene, the conditions in large parts of Holland were most favorable. In this deltaic area, rivers and rainfall provided abundant water. Temperatures were moderate under the prevailing marine climate, and drainage was poor.

When a lake is filled with plant remains—peat—it gradually is converted into land. The kinds of life that grow change in a well-defined sequence. In general, floating water plants and algae form the first debris to be accumulated in the lake. When the water is less than 6 feet deep, reeds can take over; at a depth of 20 inches, sedges dominate. Finally the ground is high enough for trees to develop, leading to a type of peat full of roots and stems.

These successive plant communities all depend on ground water for their development, and cannot grow much above the gound-water level. With enough rain, however, peat moss or sphagnum may dominate the scene. It has a very peculiar structure which allows it to hold water. In hot dry periods it uses up its water reservoir and appears brown and dead. But during a shower it sucks up large quantities of water, and appears green and healthy again. Sphagnum also recycles its food very efficiently because it can grow high above the ground-water level, fed only by the nutrient-poor rainwater. This peat moss may rapidly outgrow and suffocate the trees and shrubs that were forming the wood peat. As it does it can produce mossy cushions up to 20 feet high covering hundreds of square miles. These are curious constructions: gigantic water mattresses, pervaded and kept in place by a fine network of organic remains, and forged by a thin veneer of living, teeming tissue on the surface.

This is what happened on a huge scale in Holland. The

principle of the distribution of the different types of peat is shown in figure 1.4. The zones bordering the rivers and streams were regularly flooded and received a good share of nutrients and clay. Under these conditions, wood peat dominated. Close to the river mouths, where the sea turned the water brackish, reed peats were laid down, while sphagnum cushions developed in the large areas in between. This rather astounding development of peats illustrates well the role that life has played in this area during the past few thousand years. There are few other geological forces that rival life in raising, by several yards, a stretch of land of this size in such a relatively short period of time.

Some 1000 years ago, the terrain around Nieuwkoop was wilderness. The Meije was just one of many streams in the region that removed superfluous waters from the peatlands. Away from its banks, the surface gradually rose and was covered with swamp woods. Then, at about the present location of the open lakes, a huge sphagnum cushion with very few trees started to form, just as we would expect. The contrast with the present is dramatic (see fig. 1.5). The only thing that seems the same is the course of the Meije. How did the present situation evolve from the earlier, natural one?

The Impact of Culture

The wilderness was a forbidding place; even the Romans avoided it. At the beginning of the thirteenth century, however, increasing population made exploitation inevitable. The area was brought under feudal control of the Counts of Holland and the Bishop of Utrecht, and a very methodical cultivation system was initiated.

Drainage is the major problem for cultivation in this

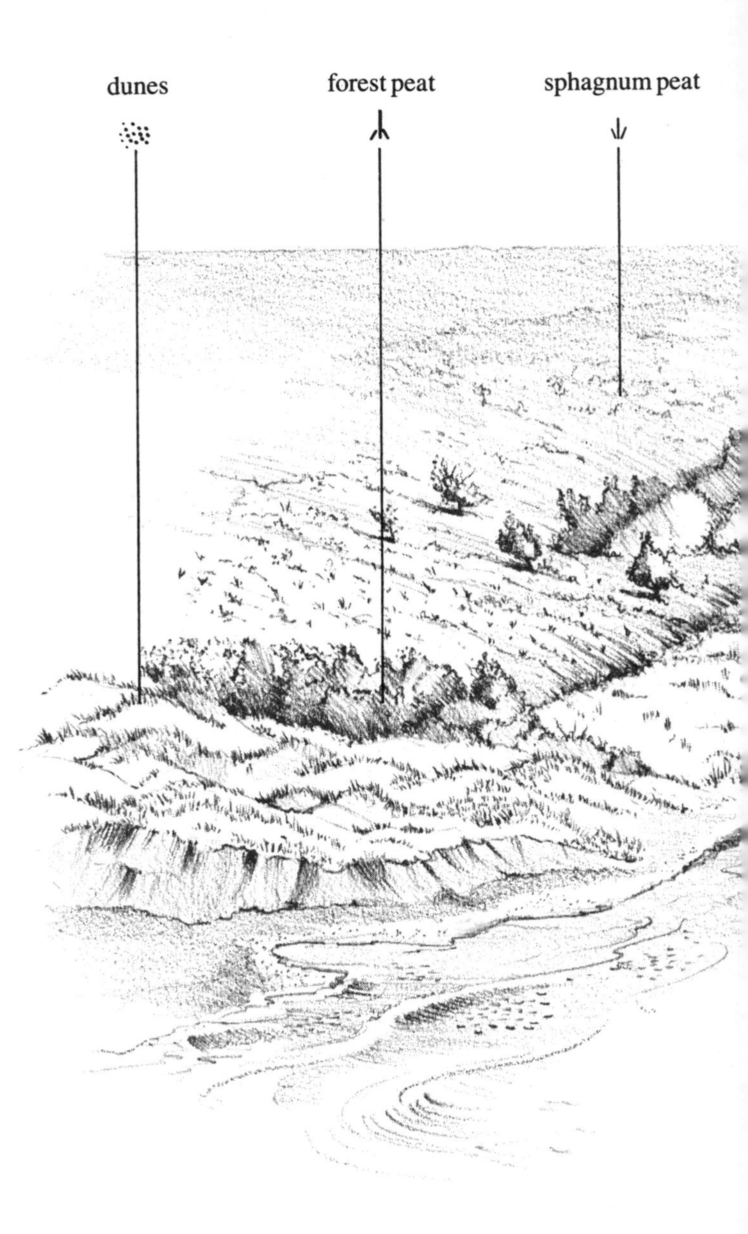
dunes
forest peat
sphagnum peat

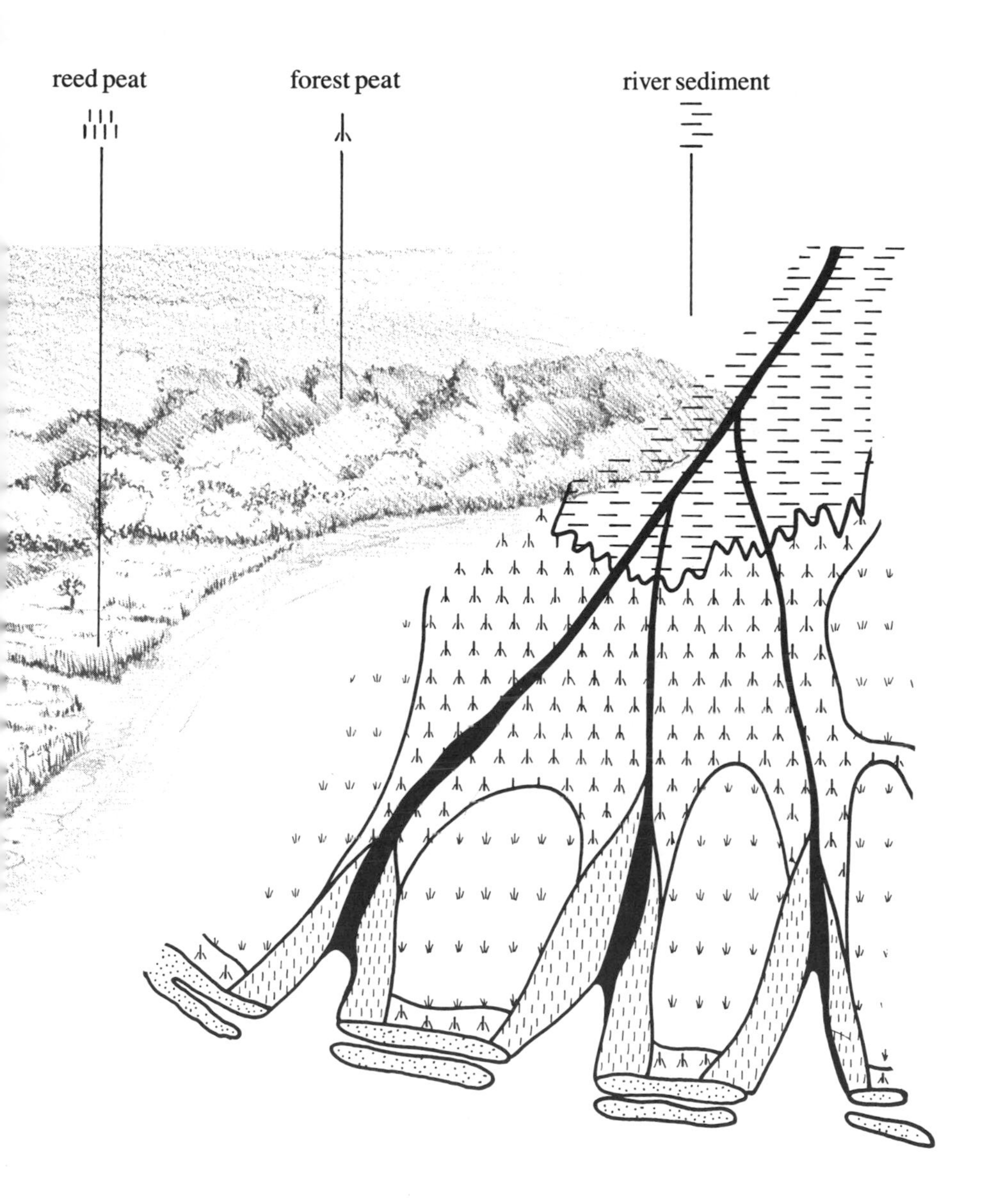

1.4 Distribution of various types of peat in a deltaic area such as Holland.

region. But at Nieuwkoop, at the edge of the sphagnum cushion, this was easily overcome, and it is here that local exploitation began. Farmhouses were built in a line along the edge of the cushion, and narrow plots, 100 to 200 yards wide and separated by ditches, were extended into the bog for a stipulated distance, generally 1250 to 2500 yards. Reclamation started from the farmhouses, as the first settlers dug drainage ditches to lower the water level. The peat

1.5 Cross sections through the Nieuwkoop and Meije areas 1000 years ago and at present. *Redrawn from J. Teeuwisse, De ontwikkeling van het landschap van 1000 tot 2000; in* Nieuwkoop, Beelden en Fragmenten *(Noorden: Post, 1982).*

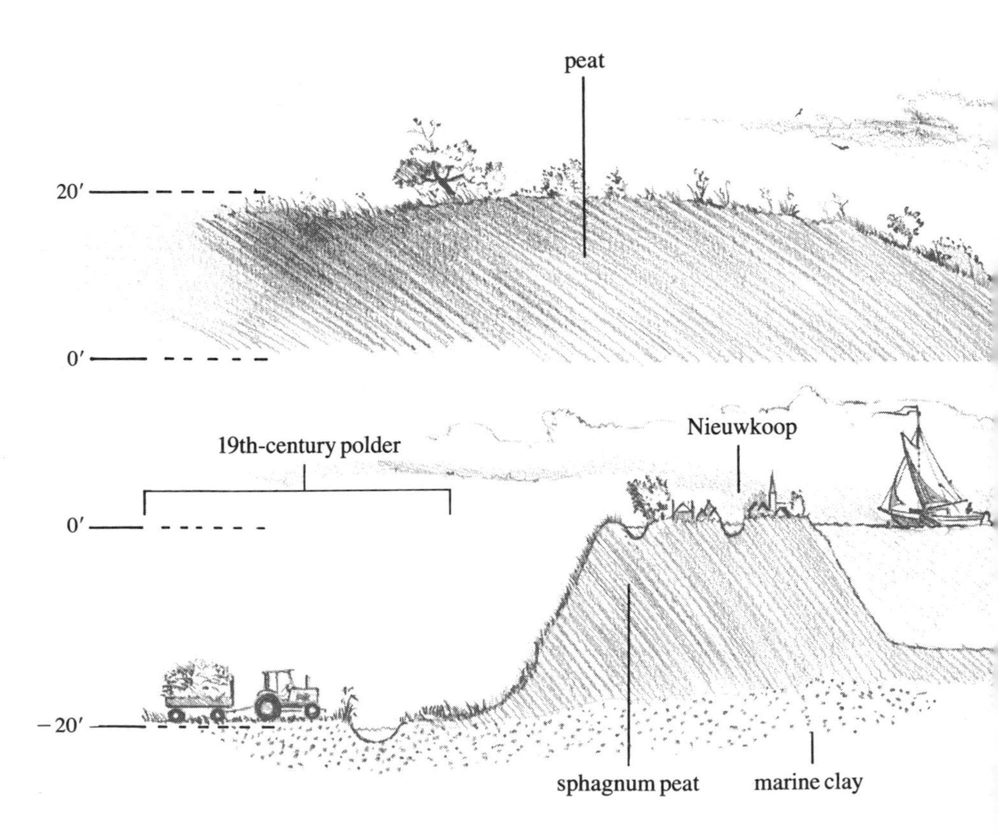

excavated from the ditches was mixed with manure and spread out over the land, and the vegetation was burned to speed the return of nutrients to the soil. The settlers cultivated grains, and kept sheep, not only for their own consumption but also to sell to the growing wool industry in nearby towns such as Leiden. More than a generation elapsed before a single row of fields was brought into cultivation. When the terminal line was reached, it became the starting

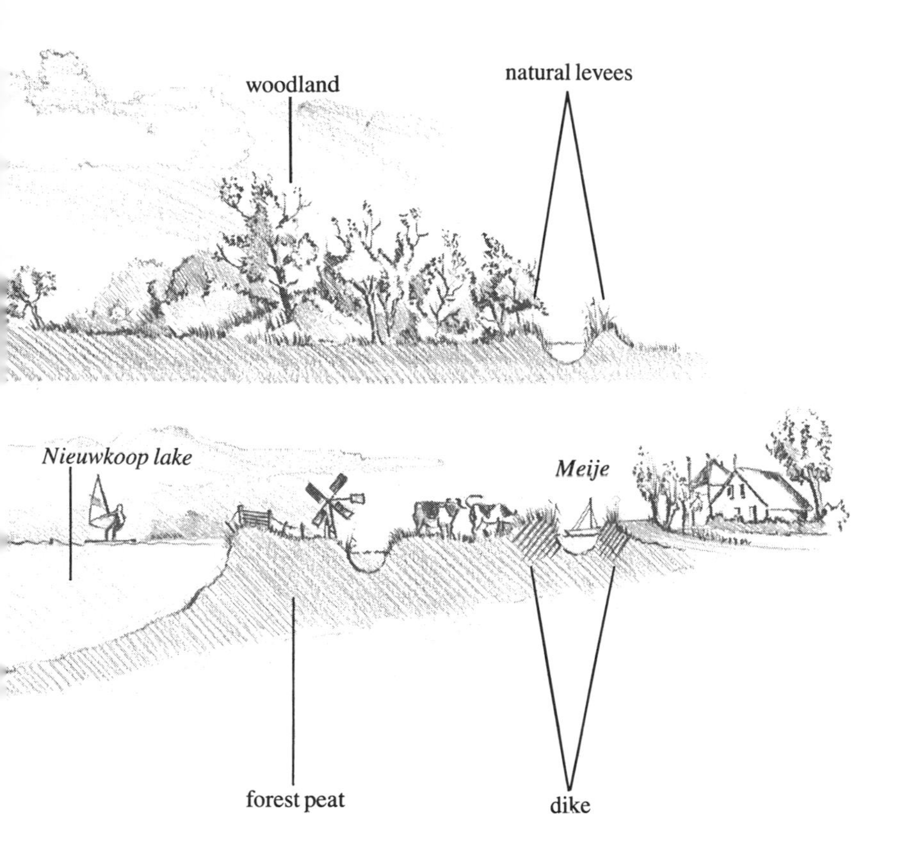

point for a second generation of exploitations. As a result, the countryside was divided into a remarkably uniform sequence of parcels. Even now, this is a very characteristic feature of the Dutch landscape—we saw it at the beginning of our trip, on either side of the Meije.

A large part of peat, in sphagnum as much as 80 percent, is water. Drainage causes a swamp to shrink, accelerated by the increased exposure to oxygen that stimulates the breakdown of plant debris. The settlers soon discovered that their activities caused their lands to subside and be drowned to an ever greater extent. They deepened the ditches several times, and removed water from the surface by manual labor. A vicious cycle began: the improved drainage caused further collapse, forcing the peasants to take more severe measures. The existing swamp streams, such as the Meije, could not remove all the excess water. Drainage works in one area brought flooding elsewhere, which caused many skirmishes between the local landlords. But the flooding also imposed the construction of an intricate system of land drainage and reclamation, and, concomitantly, it led to the creation of novel technologies in water management. A complex network of new waterways, canals, ditches, and dikes appeared in the landscape and as a result, new exploitations became feasible. By the end of the Middle Ages, the whole region around Nieuwkoop was in cultivation.

The invention that revolutionized the development of the Dutch landscape was the wind-driven water-pumping mill. The earliest version appeared in Alkmaar in 1408. It was small, driving a paddle wheel or a scoop wheel, and could carry water five to six feet upwards. A series of two to four such mills was needed to drain deeper waters. More sophisticated and much more efficient windmills using Archimedean screws superseded the older types, and by the seventeenth century were a common feature. They made it possible to drain large terrains that otherwise would have fallen victim to occasional incursions of the sea. New types of polders came into being. They were surrounded by *ring dikes* and, outside these, by ring canals. The windmills discharged their surplus water into these receptacles, to allow for transferal to the main rivers. The improved drainage allowed the polders to be used as grasslands, which made cattle farming profitable. The development of Dutch cheese went hand-in-hand with the introduction of the windmills.

One important element in the development of this landscape has not yet been mentioned. The peasants lived atop a thick peat blanket of excellent fuel. Originally, they dug away some of it to satisfy their own needs, and it was readily replenished as peat accumulation resumed. But the demand for fuel increased as the population grew. In particular, more fuel was needed to support the growing towns in a wide variety of industrial activities: breweries, potteries, metal and cement industries, brickworks, and so on.

In the early days, only a superficial layer of peat was dug away, but in 1530 an important innovation was introduced that would leave deep scars in the Dutch landscape. It was the *baggerbeugel,* a long-handled metal net which allowed peat to be dredged from several feet below the water level. Figure 1.6 shows the peat mining in operation. Large and deep rectangular pools were excavated, and the peat was spread out as a slurry on narrow strips of land in between, where it was dried and carved into blocks. The complex labyrinth of land ridges and waters near Nieuwkoop is the remnant of just such an exploitation. One reason for the resounding success of peat mining was that the fuel could be transported very cheaply by boats over the elaborate system of canals. Peat supplied the energy that stoked the industrial centers of the seventeenth century, what is known as the Golden Age in Holland. If horsepower had been needed for transport, the enormous feeding expenses would have precluded the development of this remarkably prosperous economy.

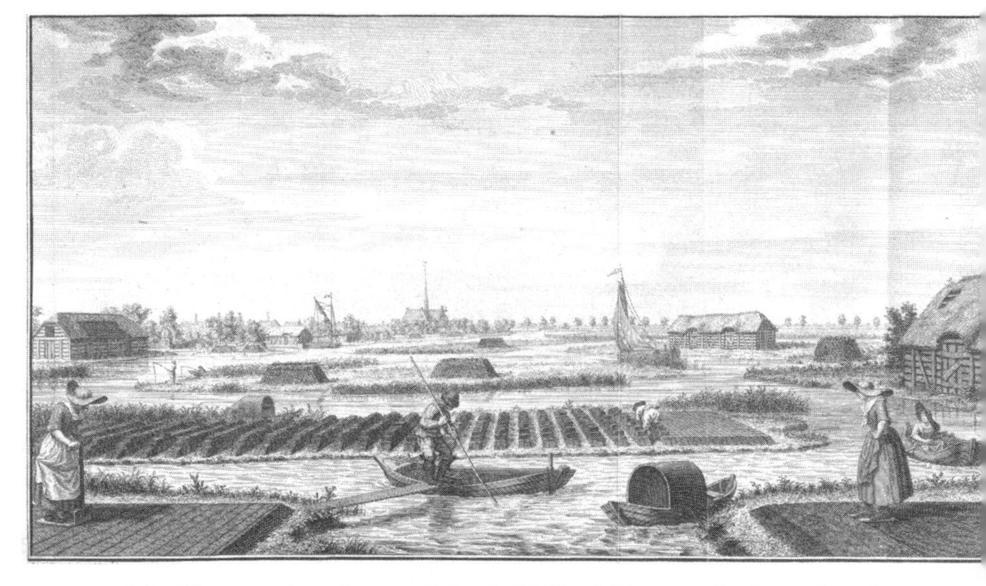

1.6 The operation of peat mining in Holland. The man in the rowboat is digging for peat with a net on a long pole: a *baggerbeugel.* The peat is spread out over elongate strips of land, dried, and prepared for transport by boat. *Engraving by J. C. Philips (1741).*

Unfortunately, the hunger for fuel in the cities of Leiden, Gouda, and Amsterdam also led to widespread destruction of the land. The skeleton of narrow landstrips left in the wake of the underwater excavations was an easy target for wave erosion in stormy weather. Poverty encouraged the rural population to sacrifice the meager long-term benefits of the land for attractive short-term profits. For the majority of people, the consequences were detrimental. Poverty increased, and depopulation of the area followed. Regulations issued by the local authorities to curb the destruction were circumvented. Large territories laboriously brought into cultivation in earlier days gave way to steadily growing lakes that could hardly be kept in check. The advent of the more powerful windmills in the seventeenth century meant that some of the land could be reclaimed as polders; the last remains of turf were sold, and the exposed marine clays were used for agriculture or grassland for cattle. But it was only in the nineteenth century, with the introduction of the steam pump, that the deterioration of the landscape could be brought under control.

Now the country is drained by an elaborate system of electrical pumps. Large polders have been reclaimed in recent years, and only a few major works remain to be carried out. Some lakes, such as the one between Nieuwkoop and the Meije, were preserved—their exploitation would not have been profitable. At one time, the government seriously considered using the Nieuwkoop lake as a dump for the garbage of the towns and cities in the area, but this plan was thwarted just in time by the nascent movement for environmental protection. The area is now a popular holiday resort.

View from the Tower

These days, the gray, stout water tower of the town of Meije stands out amidst the lush grasses and provides a vantage point for viewing the historic area. One sees the remains of an extensively excavated mining district, and at the same time, a landscape that for centuries has been the scene of careful and parsimonious cultivation and water control. Nothing here is left to chance. Even the reeds nearby have been maintained over the ages to supply the tulip fields near the coast with a protective covering during the winter months.

Human society has subjected the wilderness, once formed by the unbridled geological forces of nature, to one of the most telling examples of interference with the environment. At first this interference was destructive and brought about a dangerous imbalance in the natural equilibrium. Later, the ruinous effects were curbed by protective countermeasures and, in due time, a new human-maintained balance was achieved. This country has gone through a long succession of such cycles of destruction and reclamation. The result is a pleasurable homeland where we can enjoy the interplay of water, wind, life, and history without being overexposed to the vicissitudes of nature. The impenetrable wilderness has become a friendly landscape, and a delight for the city dweller.

Today, the integrity of Nieuwkoop's surroundings is under renewed attack. Widespread use of fertilizers is converting the nutrient-poor and species-rich fields into dull monocultures of grass. The waters, once potable and crystal clear, are being polluted with nitrate and phosphate. Sewage, as well as local industries and tourism, are additional causes of deterioration, and there are only a few places left where one can safely swim in the summer months. Birds and fish are threatened, and encroaching urbanization undermines the comfortable peace of the olden days. Measures are being taken to curb this new instability, and the local government is to be praised for its strict policy of environmental care. But will this be enough?

A new aspect has entered into the perennial cycle of

destruction and reclamation. The present tendencies are not restricted to this region alone, but operate on a huge scale in the world at large. In the past, the character of human intervention differed from place to place and was dependent on local traditions and the state of the land. In contrast, the present rapid standardization and intensification of production methods brings about an unprecedented, centralized buildup of the industrial effort. The hazard has become a global one and calls for a global response.

Such concerns have spurred a broad-based research effort known as The International Geosphere-Biosphere Programme: A Study of Global Change. The IGBP draws on the strengths of many scientific disciplines in order to describe and understand the physical, chemical, biological, and cultural processes that regulate the earth. In this way, the participants seek to discern the conditions that comprise the earth's unique life-supporting environment, the changes that are occurring in the system, and the ways in which these conditions are influenced by human actions. The benefits of this approach are many. One will be advances in our understanding of the ways individual elements of the earth system operate. Another will be improvements in our ability to forecast environmental changes, a task made all the more difficult because the impacts of human actions occur against a backdrop of natural environmental fluctuation.

In the course of conversations with my fellow scientists, I often discussed plans for the IGBP and pointed to the need for a rational balance between the forces of nature and culture. To my surprise, enthusiasm for this concept was not general. Several colleagues felt that any further human interference with nature could lead only to disaster;

they thought the idea of global management was a dangerous and misguided illusion. I cannot help thinking that our feelings on these matters are related to our cultural heritage and our personal experience with the interaction of nature and culture. For a Dutchman like myself, born and bred in this muddy polderland, the idea of a cultivated earth is not illusory, but a matter of common survival.

The geological effects of the physical, chemical, biological, and cultural forces can never be separated completely, and it is impossible to evaluate and quantify their respective influences in detail. Where are the specific physical, chemical, biological, and cultural elements in Holland's waters, skies, vegetation, houses, ditches, and dikes? These forces form a matrix, the divisions of which exist only in our minds.

How curious our planet is. The moon, Venus, Mars, Jupiter . . . they are neither dead nor alive, they just *are*—senseless, indifferent. The earth is the only place in the solar system where life and death could emerge; where things are organized, acquire a sense or a meaning, and are then disturbed and destroyed; where abundant beauty and harmony are interwoven with tragedy, suffering, and ugliness.

Those who founded the sciences of life and earth 150 or more years ago intuitively adopted an integrated approach to the study of nature. Subsequent specialization has brought great advances, but has also obscured our view of the unique role that life and death play on our planet. This segregation is inhibiting research. We must bring sciences together so that an integrated earth science can be reconstructed.

PART I

Geology of the Planet Earth

2

SHIFTING PLATES

WHEN HANNIBAL was asked what he saw when he crossed the Alps, he replied, "Just a heap of rocks." The tourist of today has less prosaic tastes. The rough scenery is there to be enjoyed. The lavish pastures and forests and the wild mountain streams remind him of a pristine and unspoiled world.

Geologists have yet another appreciation. While strolling through the mountains they perceive the deep sea of an ancient past, lagoons and reefs, beaches and lakes. Or Hadean depths not normally accessible to humans, with temperatures so high and pressures so great that rocks become plastic, fold, melt, and resolidify. A geologist will show you proof of giant forces that break up the crust and transport the fragments over hundreds or thousands of miles, uplift entire mountain chains, and force vast masses to subside into the depths. A geologist's perception is far from naive. He selectively scrutinizes the environment and is able to recognize a few key phenomena that escape the layman's attention, but provide information about the earth's history. His guide is geological theory, a body of thought that emerged during centuries of painstaking research. Bit by bit, geologists build a picture of a succession of past worlds, intangible, of unimaginable duration, yet as real as the rocks themselves.

Over the last few decades, the geological perception of the world has changed dramatically. We can now perceive things on a much larger scale than before. It is as if we had moved into space and were viewing our planet from a distance and with a condensed perception of time. Regional histories fall into place as parts of a jigsaw puzzle, and giant events become apparent that affected the earth as a whole.

Plates

Crucial in this development was the advent of the plate-tectonic theory in the early sixties. No aspect of geologic research remained unaffected. Volcanism, earthquakes, the relief of the ocean floors and continents are all now revealed as part of an all-embracing master scheme. We understand how the structure of Java or Japan is related to the history of ancient mountain belts, or to intricate patterns of life's evolution in the distant past.

The earth's surface is not a continuous global sheet, even though it seems to be. Rather, it is made up of huge crustal plates that move and shift. The lithosphere, of which the plates consist, includes not only the earth's crust of oceans and continents, but also the top rigid layer of the earth's mantle. Underlying the lithosphere is the plastic asthenosphere, a deeper zone in the mantle over which the lithospheric plates can slide.

Figure 2.1 sums up the principle of plate tectonics. It shows the Atlantic Ocean in the middle, and Africa and South America, with part of the Pacific on either side. The Atlantic is the scene of spreading: two plates slide apart carrying North and South America on the one side and Europe and Africa on the other. The boundary is marked by the Mid-Atlantic ridge where molten rock erupts from the asthenosphere to fill the gap left by the moving plates. The molten material solidifies into basalt, forming new ocean

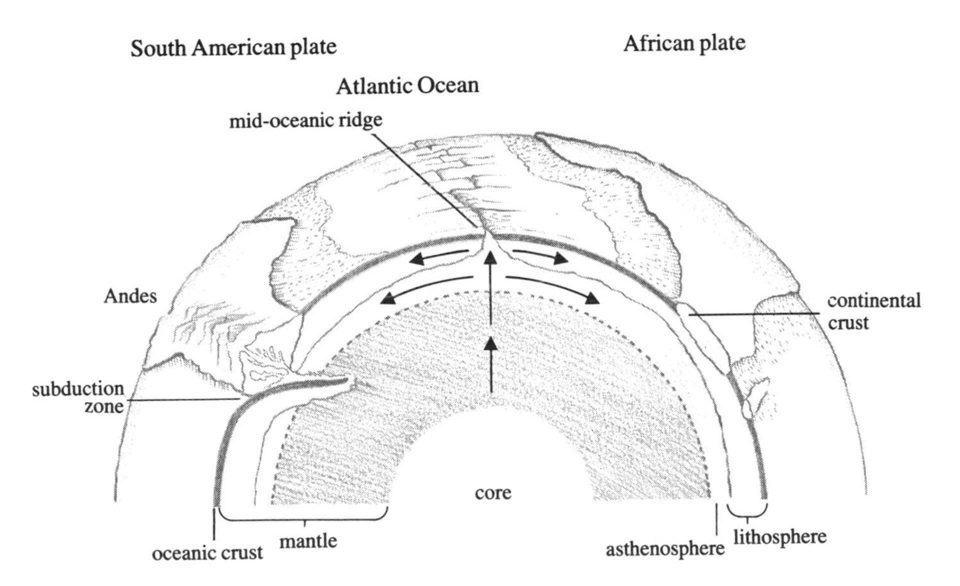

2.1 Principle of plate tectonics. *Adapted from* Earth, *3/E, by Frank Press and Raymond Siever. Copyright © 1974, 1978, 1982 by W. H. Freeman and Company. Reprinted with permission.*

crust that moves with the plates. The youngest material is always nearest the mid-oceanic ridge, and the age of the ocean floor increases equally in both directions away from the ridge.

If fresh lithospheric material is created in one place it must be destroyed in another, or the globe would be an ever-expanding balloon. In contrast to the Atlantic, the Pacific is a shrinking ocean, surrounded by subduction zones: plate margins where the ocean crust is being subducted, or drawn into the mantle and consumed. The example shown in figure 2.1 is the subduction zone bordering the Andes. Here, the Pacific plate and the crustal plate carrying South America move toward each other. The Pacific plate is pushed beneath the lighter continental plate, and a deep trench forms along this line of contact. It is a zone of friction and earthquakes. The overridden rock slab is heated, melted, and partly absorbed into the mantle. Lightweight components, however, escape upwards to the land surface. They form a stream of rising magmas and lavas that are associated with mountain building and extensive volcanism.

The motion of the plates is driven by heat released by radioactive decay deep inside the earth. Within the mantle, the molten rock expands through this heating; it becomes lighter as a result and rises toward the surface. There it is pushed aside, cools, becomes denser, and eventually is drawn back into the earth. When the rocks sink, they are warmed again. The whole system operates on the same principle as a central heating system in a Roman villa: convection currents. The complex system of these circular flows in the earth's mantle drags along the lithospheric plates.

The recognition of these global patterns of movement has not only revolutionized our understanding of the earth as a whole, but is also the key to regional studies. I give two examples. The first is an area where the geotectonic forces are still in full swing. This then serves as an introduction to the other, more difficult case in which these motions came to a halt long ago, so that only their petrified vestiges are on display in the field.

Java

Java is an island in the making, a narrow strip of land stretching along the boundary between two shifting lithospheric plates. The daily life of its people is threatened by earthquakes and volcanic eruptions. Only thirty years ago, it was not possible to relate these devastating events to the nature and distribution of rocks in the Javanese region or to the curious landscape of the sea floor around. Geological reconstructions were kaleidoscopic, fragmentary. But now, with plate tectonics, the parts can be melded into a single coherent reconstruction. In the Java area the process of subduction is in full tilt: the ocean crust, formed long ago at a mid-oceanic ridge, moves towards a neighboring plate that carries the island, and then plunges into the earth at a rate of a few inches per year.

The various processes associated with subduction are clearly expressed in the general topography of the area. These regimes are arranged in zones parallel to the length of the island (fig. 2.2). Starting in the Indian Ocean, the sea floor deepens to form the Java Trench, some 150 miles south of Java. The sea floor then begins to rise, culminating in the Java Ridge less than 3000 feet below sea level. Between the Java Ridge and the main island is a depression, a long basin about 100 miles wide and with water up to 12,000 feet deep. Then comes the main volcanic island and, to the north, the shallow Java Sea, no more than a few hundred feet deep. So there is a succession of five zones from south to north: the deep sea with its bordering trench, the Java Ridge, the basin, Java itself, and the Java Sea.

The crust under oceans is only about 3 miles thick and consists largely of solidified magma of basaltic composition—dark and rather heavy rock. A sheet of lightweight sediments—basically detritus derived from the continents—is accumulated on top. Because of its low density, this sediment cover tends to remain "afloat" during subduction. Much of it is scraped off by the overriding plate. It is crumpled, crushed, pushed up, and telescoped together into a deformed mass of rocks bordering the trench. This is how the underwater Java Ridge formed, a process still underway.

On its way down, the descending slab of the Indian Ocean floor is warmed up until, at depths between 60 and 120 miles, some of its minerals melt. The blobs of melt coalesce and form bodies of magma that move upward. They intrude into and partly melt the overlying rocks, and then resolidify. As the heat flow continues, a new cycle of partial melting, ascent, and solidification begins. Finally, some of the magma may escape to the surface, giving rise to volcanism.

During this process the composition of the magma changes dramatically. Instead of the dark, heavy basalts of the deep sea, lightweight rocks closer in composition to granite are generated. Java is an igneous plug capped with volcanic deposits, a lightweight slag of rock exuded steadily from the subduction zone. This is a major mechanism by which new continental masses are created.

Thus we now understand the significance of the deep ocean trench, the Java Ridge, and the island of Java for this highly dynamic scene. The 100-mile wide basin between the ridge and the island marks the zone where the ocean plate dives toward the mantle but is not heated enough to produce any volcanic activity. It simply acts as a trap for debris eroded from the steadily rising volcanic arc. Giant forces are at work here: in this depression a pile of sediments more than 15,000 feet thick has accumulated below the 12,000 feet of water.

Indeed, Java beautifully exemplifies how plate tectonics is a key not only to the dynamism of the earth as a whole, but also to geology at the regional scale. We immediately understand that the volcanism, the earthquakes, the trench, the underwater ridge, and the basin behind it as well as the nature of the rocks on the main island are aspects of a single process: subduction.

And now we come to the difficult part. In a mountain belt formed long ago, faint traces of the earth's spasms way back in the past can still be perceived in the rocks. The

2.2 Java: example of a subduction zone.

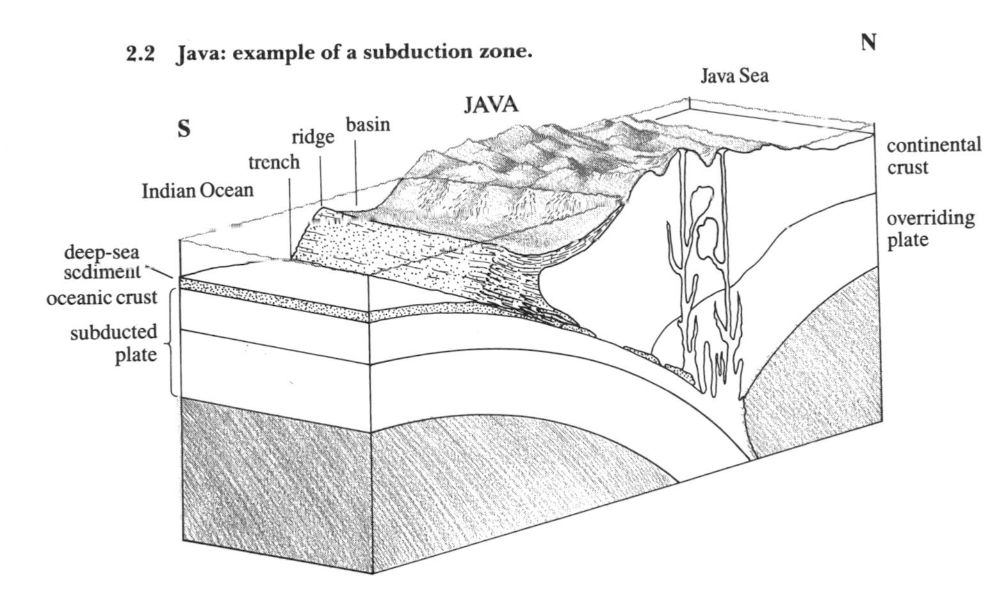

evidence is distorted and covered with the dust of the ages. We must carefully assemble the bits and pieces until the whole picture emerges. Subduction, the concept that explains the present dynamics of Java, will be our guide to the past.

The Fossils' Tale

A small, green slice of Scotland reflects the many facets of scientific inquiry required to unravel the area's complex geology. The story began in 1962 when Alwyn Williams, then professor of geology at Queen's University in Belfast, completed a thorough geological study of the Girvan district about 50 miles southwest of Glasgow (fig. 2.3). Williams's work was limited, in geographical scope and also in time, to rocks that formed some 450 to 480 million years ago, during a period known as the Ordovician. However obscure this subject seems, the investigation became a classic. Through painstaking research combining diverse geological disciplines, Williams reconstructed the area's dramatic past. That interpretation eventually formed a cornerstone of our present understanding of the whole of Scotland and its geological setting at large.

One can review the story Williams discovered by driving through the Girvan area. The road trends east along the Stinchar Valley, away from the Irish Sea, toward the old town of Barr. On a sunny Sunday morning in June, a cool breeze blows a few scattered clouds over the fields. The countryside is green, glaringly green. Green grass covers the hills, and green trees line the valley.

The hills that border the south side of the Stinchar Valley consist of masses of marine sand and clay deposited more than 400 million years ago during the Ordovician (fig.

2.4). On the valley's north side, the rock outcrops in the hills are huge deposits of conglomerate—solidified heaps of boulders and pebbles. Despite their differences, the rocks on either side of the valley are the same age. In other words,

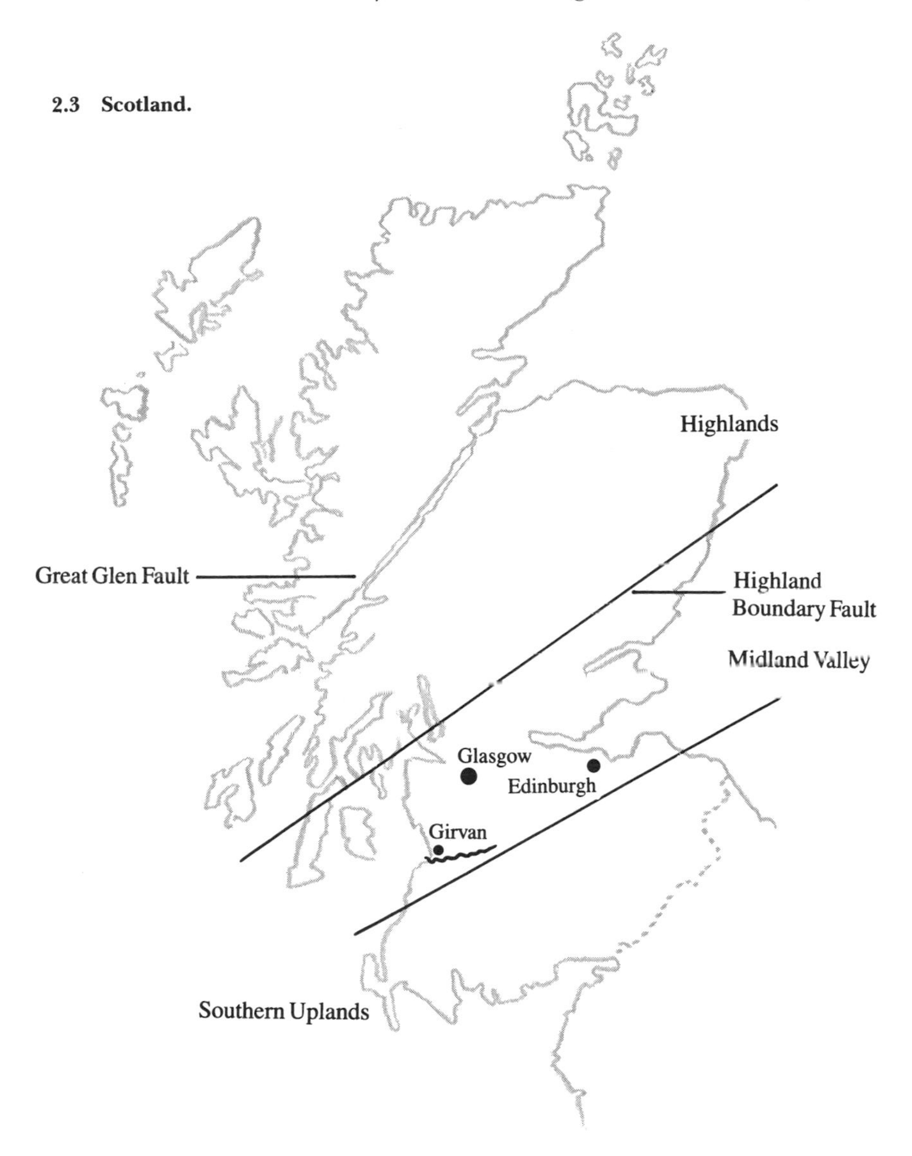

2.3 Scotland.

coarse debris was deposited on one side, while finer particles were laid down on the other. Occasionally, a tongue of conglomerate protrudes into the sand and clay sequence.

The valley is at most a few million years old. It formed when the Stinchar River cut its way through these ancient rocks. To learn how the rocks were formed, one must ignore the valley and imagine that the rocks on either side were a continuous mass, as in the cross section shown in figure 2.5 (A). Here one confronts the perennial problems of geological research: to see the dynamism underlying the

2.4 Stinchar Valley.

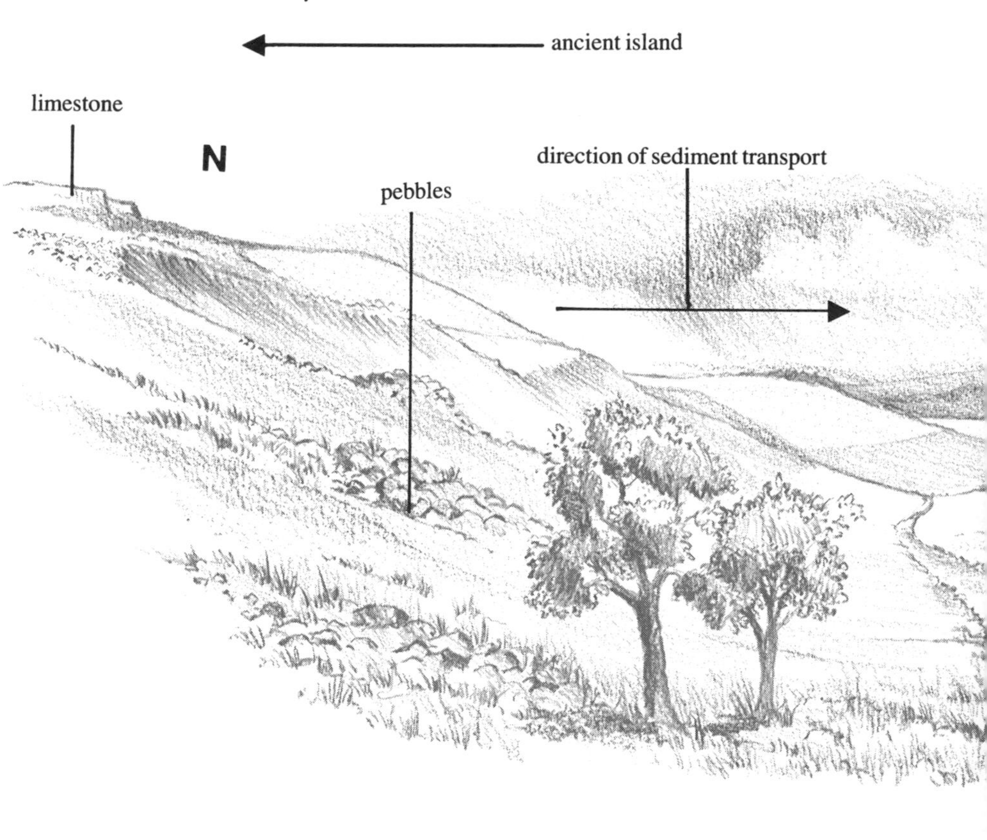

rigid structure, and to reconstruct the genesis of the strata. As a rule, the best approach is to make a detour and look for a similar situation in the recent environment.

One would start by looking at the accumulations of boulders and pebbles. These form at the foot of mountains; they are pieces of rock, detached by weathering. Thus, the Girvan District must have been a mountainous area when the rocks were formed. Where exactly were these mountains? At the foot of a mountain chain one sees how the weathered rocks are washed downslope by the rivers. The

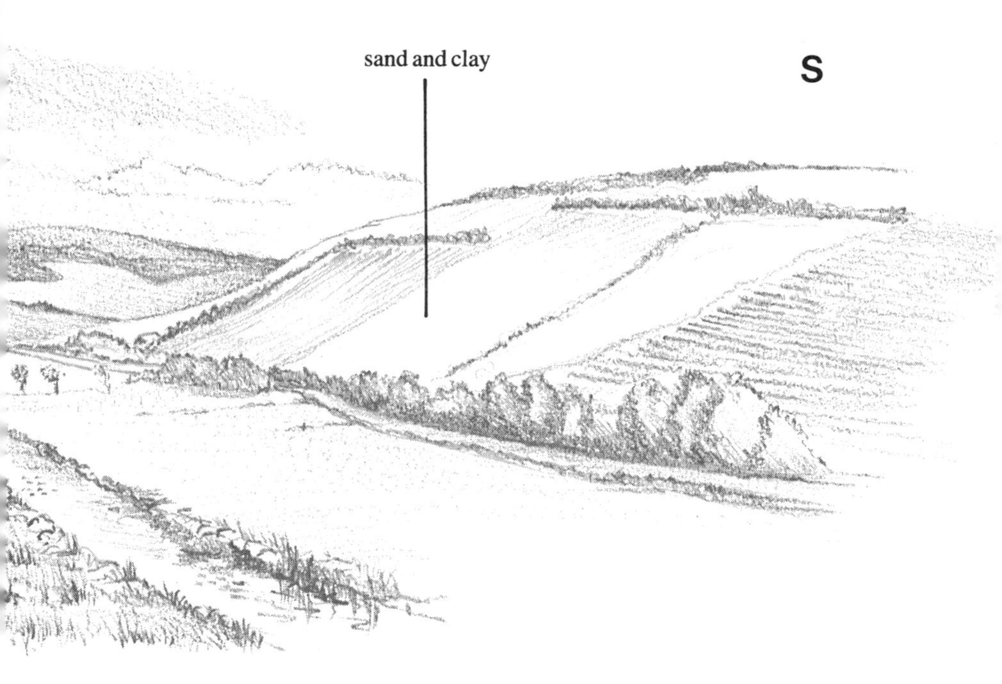

debris is sorted out during this transport. Boulders and pebbles are not easily transported, so they remain close to the place where they were detached from the rocks. Sand and clay, on the other hand, are dragged along with the water and accumulate downstream. One may conclude that when the Girvan rocks were formed, the mountains were right behind the conglomerates [fig. 2.5 (B)].

My companion, Brian Bluck, a geologist at Glasgow University, parks the car near a farmhouse, and we set off to climb the hill north of the road, towards the Ordovician mountains, to have a closer look at the conglomerates. The exposure is in a ditch, below the roots of a tree. With our hammers we chip pieces from boulders and look at them through our hand lenses. The chips are granite, an igneous rock formed when magma cooled deep in the continental crust.

Higher up the hill is an outcrop of limestone. It is an ancient reef formed of algal remains and shells of organisms, without any land-derived pebbles, sand, or even clay. The interpretation of this rock requires comparison with a totally different scenario. Today, similar limestones form far away from the mountains in shallow seas, such as Florida Bay. The conclusion is straightforward: the mountains rose up from the sea, and it is then that the debris was shed. This process continued until the mountains were eroded away and the sea was nearly filled with rock. The influx of debris stopped, and conditions then favored the formation of limestone [fig. 2.5 (C)].

Time after time the land was uplifted and then eroded. The fact that the conglomerate pebbles consist of granite shows that the uplifts were caused by intrusions of rising granite magma. Convulsive events were separated by long, quiet periods during which reefs could grow in the shallow

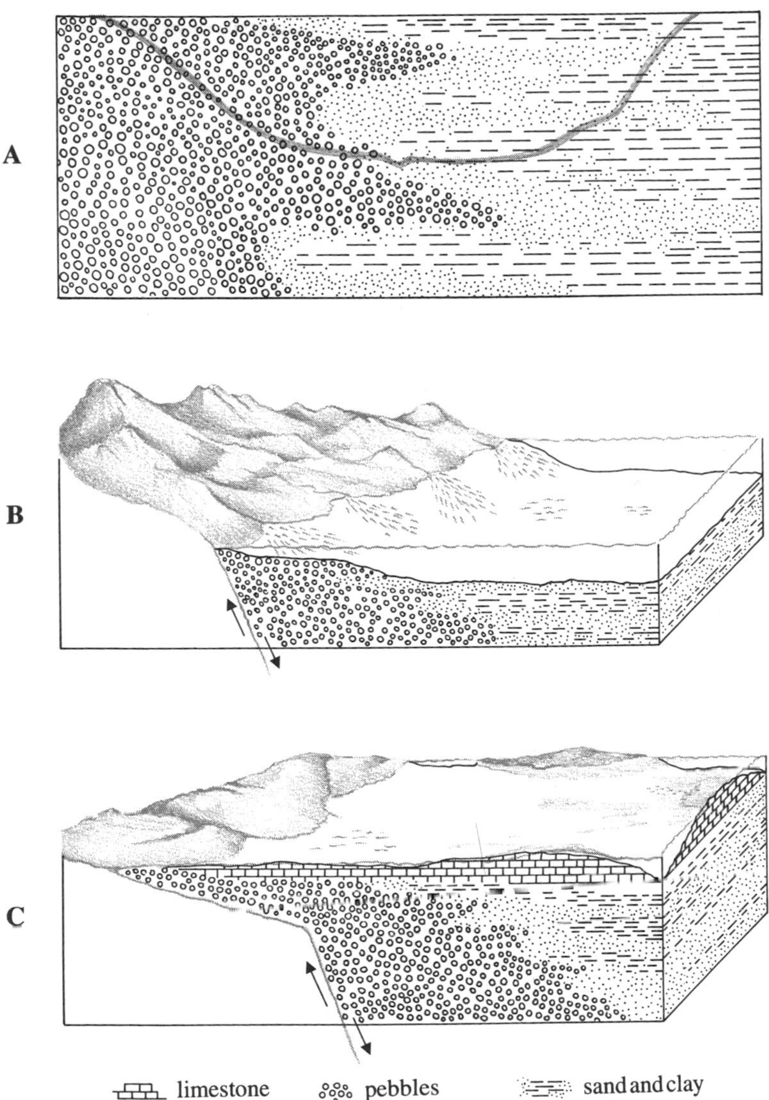

2.5 (A) Position of pebbles, sands, and clay in cross section. The present valley is a late feature, and cuts through the rocks. (B) Dynamical interpretation of figure 2.5 (A). The sediments are thought to be erosion products of an uprising landmass on the left (north). (C) Erosion has leveled off the landmass after its uplift has come to a standstill. The transport of pebbles, sand, and clay is suspended, and limestone reefs can form in the shallow sea.

waters topping broad conglomerate platforms.

Figure 2.6 shows Alwyn Williams's reconstruction of the area. It reveals a stepwise succession of blocks, separated by faults, which rises from south to north. This part of Scotland is part of a landmass that emerged from below the ocean floor and built up to the surface in a series of spasmodic emanations of magma. The work reflected in the figure is particularly valuable because it is supported by a detailed body of paleontological evidence. Williams's publication contains descriptions of 180 species of extinct shelly animals, all collected from the rocks shown in figure 2.6.

The fossils belonged to the brachiopods, a group of animals now close to extinction but very common in the Ordovician when these Girvan rocks were formed. Superficially brachiopods look like the clams and oysters of today, but biologically they are very different. Most biological species live only for a short period of time by geological stan-

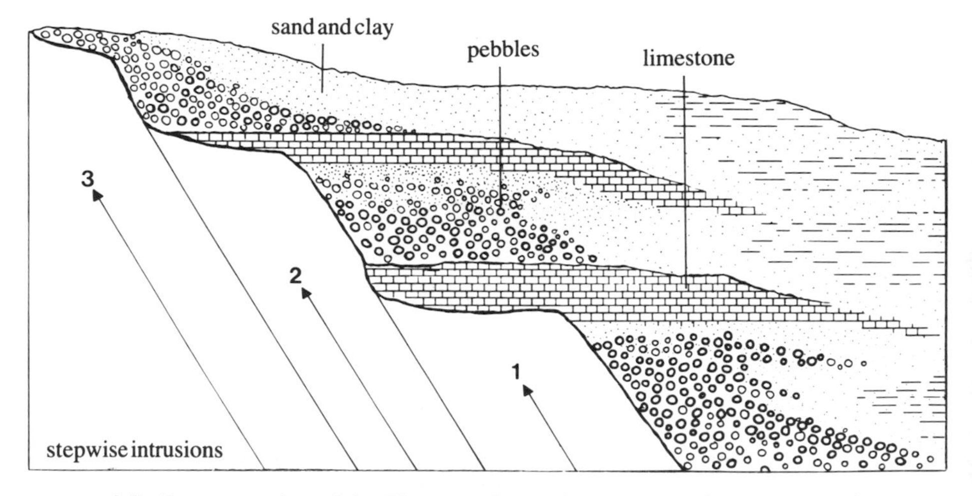

2.6 Reconstruction of the Girvan geology. The sequence of events shown in figures 2.5 (B) and (C) is repeated several times due to successive periods of uplift and stability. *Courtesy of Alwyn Williams.*

dards, perhaps a couple of million years, so their distribution in the rocks may be used as an indication of age. These fossils allowed Williams to distinguish between limestones, group them into the correct chronological order, and correlate them between areas separated by ancient faults or modern valleys. From this he could disentangle the intricate structure of the distorted sediment pile.

Another even more fascinating outcome of Williams's paleontological analysis concerned the large-scale geography in Ordovician times. Following suggestions made a century earlier by J. W. Salter, Williams's descriptions showed beyond doubt that the Ordovician brachiopods from Scotland were more similar to those from eastern America than to nearby English or Welsh brachiopods of the same age. Explanation of this curious finding had to await the advent of plate tectonics.

Iapetus

In 1966 the Canadian geologist Tuzo Wilson pointed out that during the Ordovician period, when the Girvan rocks were formed, the Atlantic Ocean did not yet exist. Instead, another ocean separated the continental masses of Europe and North America. Wilson named this ancient ocean Iapetus, after the father of Atlas from whom the Atlantic takes its name. After the Ordovician and well before the Permian period, about 220 million years ago, Iapetus had closed, bringing all the continents into one huge cluster.

Here, then, lies the answer. The fossil brachiopods of the Girvan district and England differ so because Scotland was not part of Europe, as it is now, but part of North America. Long after the Ordovician, when Iapetus closed and Europe and America collided, a huge mountain chain formed along the junction. Scotland is part of this chain and when we fit the pieces together we can see its prolongation in Norway and Greenland to the north, and in Ireland, Newfoundland, the Appalachians, and Mauritania in Africa to the south. This stimulating concept immediately triggered a host of questions. The only mechanism known by which an ocean can close is subduction, so the problem was to find vestiges of the old subduction zone along the England-Scotland suture. It was at this point that Stuart McKerrow from Oxford made a major discovery. In the peaceful countryside of southern Scotland he found the scene of the old drama: the place where the ancient ocean crust had plunged below the continent and vanished into the deep.

Understanding Scotland

Brian Bluck pulls a map of Scotland (see fig. 2.3) from his pocket and points to the Southern Uplands, a rough, almost mountainous area just north of England. Then comes a major fault and beyond that a zone with lower and gently sloping hills, the Midland Valley. We are sitting here, near Girvan, in the southwestern corner of the valley, which extends beyond Glasgow. Farther north one returns to wild terrain, the famous rugged Highlands. Between the Highlands and the Midland Valley is a major fault zone. So we have three roughly parallel areas: Southern Uplands, Midland Valley, and The Highlands. The Great Glen Fault, the spectacular northeast-southwest suture containing Loch Ness and traversing The Highlands, is in fact a minor feature compared with the other faults.

Stuart McKerrow quickly recognized the similarities between the Southern Uplands and submarine ridges such as the one south of Java. He had discovered the line along which the floor of the Iapetus Ocean had been subducted under Scotland, and believed that the suture where the former European and American continents met when the Iapetus Ocean closed must have been just south of the area where the Uplands are now. The trench had been at the border between England and Scotland, and the Southern Uplands represents the underwater ridge.

Our perch on the hilltop must be on the south flank of the ancient island arc pushed up out of the ocean, and so the Midland Valley may well be the equivalent of Java. The granite boulders and the large slope with sands, clays, and other debris across the valley are analogous to the 12,000 foot basin south of Java, filled with thousands of feet of rubble shed by the uplifting island.

"Now look at those hills at the horizon," Brian says (fig. 2.7). "Those are the Southern Uplands, the underwater ridge."

But that's only an hour's walk from here, I realize. Isn't the basin south of Java 150 miles wide?

That whole stretch of ocean floor has disappeared, Brian explains. We are sitting on all that is left of it. A little chip of sediments, just beside the island arc. One can now imagine what must have happened when the European and North American continents collided. This whole area was crushed and telescoped together. Not only was the Iapetus Ocean closed, but the entire basin between the land and the underwater ridge was removed in the process. Where it went nobody knows. Perhaps it was forced into the depths when the underwater ridge and the island were pushed together.

More likely, we now believe, the basin moved sideways, to the east or west.

As for The Highlands in the north, we now know that in Ordovician times they were nowhere near their present position. The equivalent of the Java Sea has disappeared, removed by the same forces that closed the Iapetus and telescoped the basin south of the Girvan District together. The Highlands were shipped along by plate motions from some unknown but distant place until they docked in their present position (fig. 2.8).

2.7 A southern view of Stinchar Valley, just south of Girvan. The equivalent of Java is just behind, the Southern Uplands represents the underwater "ridge," and the equivalent of the basin south of Java is squeezed out. The few hills in between are all that remains of the basin in the Scottish landscape.

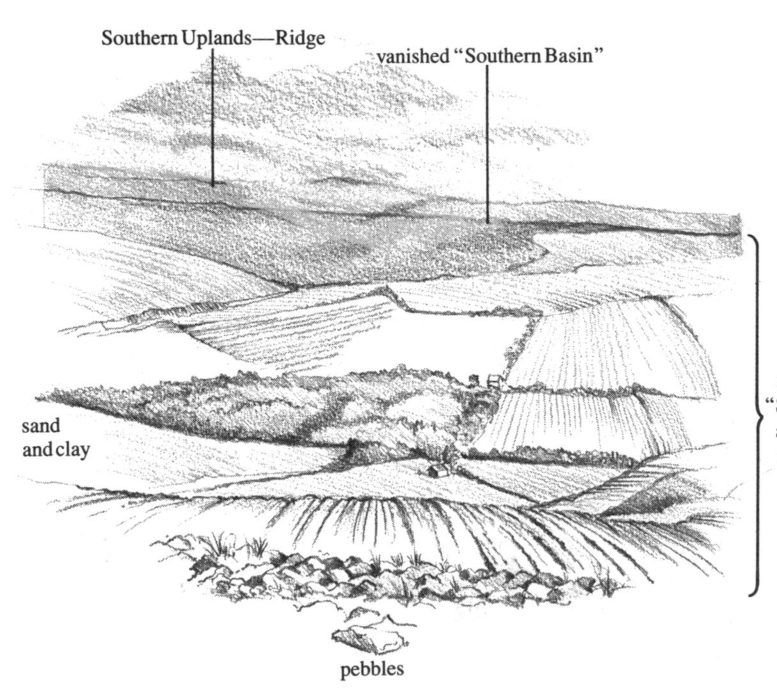

Global and Non-Global Sciences

The examples in this chapter clearly show that geologists are able to look at the whole earth and interpret the details from that global vantage point. It is good to recall here that it is through plate tectonics that geology has overcome the old, more parochial perspective. This theory makes us aware of processes that can only be fully understood when the worldwide scale of their operations is taken into account. We realize how the crust and mantle of this globe are stirred around through geological time. And this process has crucial repercussions for our understanding of the regional details, as the examples of Java and Scotland clearly demonstrate.

But there are also limitations to this newly acquired global perspective. In Scotland, we now recognize a process of subduction and continental amalgamation that took place in the remote past. An ocean was closed, and the ancient continental margins were compressed. Molten rocks were chemically differentiated in the deep earth, and their lightweight components rose to the surface to form a new landmass. Its debris was shed into a neighboring basin. The accumulated sediments were compressed and distorted, elevated here and subsided there. Throughout, our reconstruction gives little more than a probable sequence of physical and chemical events. This reflects pretty well the present state of geological theory: because the physical and chemical forces affecting our planet are the ones most accessible to investigation, the earth's development is overwhelmingly interpreted in terms of these inanimate processes. First, the physical and chemical framework for the reconstruction has to be sketched; only then can a more refined and colorful picture be painted.

Suppose that we could walk along the Ordovician shores of the Iapetus Ocean. We see, as the abundant fossils attest, that these waters teem with aquatic organisms. Microbial communities abound, reefs fringe the continents in many places, and small animals flourish. What we don't know is whether these life forms influence the great succession of events. Was life, too, a geological force like plate tectonics, volcanism, and running water? The real story may have

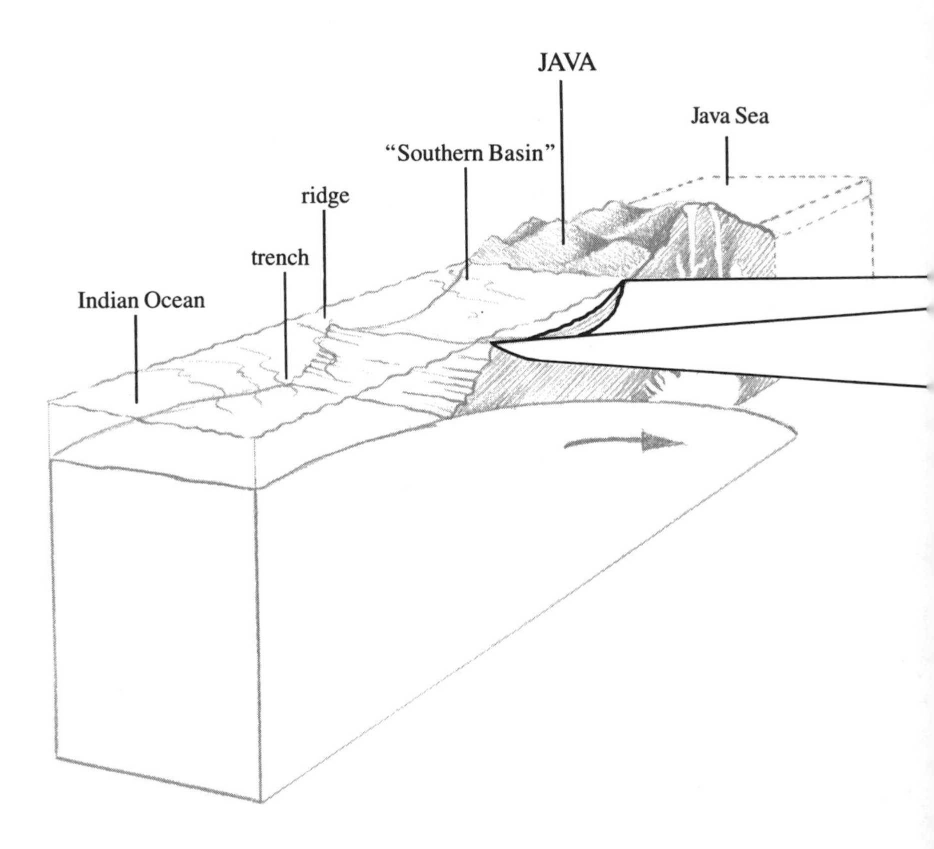

been infinitely more complex and fascinating than we are presently able to see from the rocks.

Fossils are our most valuable clues to the history of life on earth. We collect, describe, classify, and store them. Above all, we cherish them as priceless tools for geological interpretation. They give an impression of what organisms were around long ago. They also may tell us the age of the rocks, and provide information about the distribution of bygone

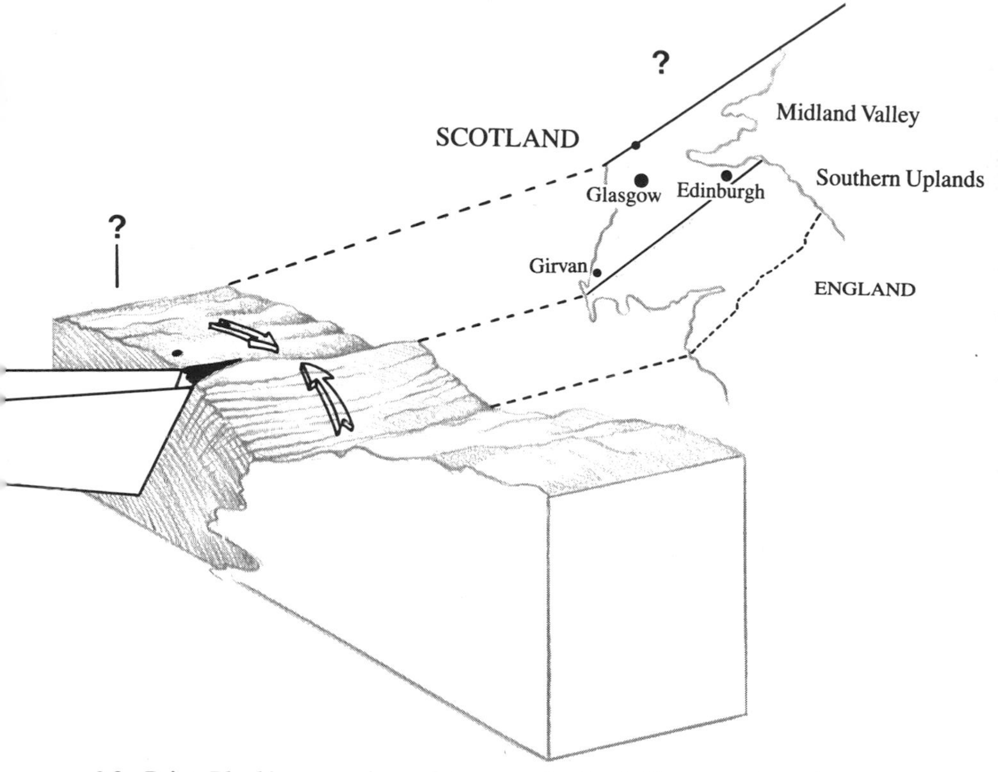

2.8 Brian Bluck's comparison of Java and Scotland. The equivalent of the basin just south of Java is telescoped. The Highlands in the north (left out of this figure) form an unrelated continental block, which slid from an unknown location to its present position along the Highland Boundary Fault (compare with fig. 2.3).

oceans and ancient environments.

Nevertheless, geological theory fails us here: it does not allow us to recognize faithfully the forces ancient organisms may have unleashed. Overwhelmingly, geologists show that inanimate forces determined the course of earth history. They consider life a decoration, beautiful but unimportant, passively adapting to the whims of the physical and chemical environment, unable to change the course of events. Clearly, life is not yet included in the geological science of the globe, while the thrust of biological research is directed towards small-scale phenomena ranging from molecules to ecosystems. What we miss is an astronomical view of life.

3

STABILITY AND CHANGE

IN THE FAMOUS wine-growing district east of Bordeaux in southwestern France, the Dordogne meanders through an undulating landscape of serene and elegant beauty. The river bends wide toward the ocean, crossing the broad valley of its own making. Viewed from a limestone precipice, the valley presents a mosaic of vineyards and grainfields, interrupted by trees, the pale reddish roofs of farmhouses, and a few villages. Low, wooded hills frame the horizon. This countryside has been inhabited by humans since the Paleolithic Age, shortly after use of tools began. Nothing is haphazard; every detail has meaning.

This lovely portion of the Dordogne Valley and its surroundings has become part of a great geological controversy. The limestone exposed in the cliffs is 30 million years old; it is part of a hard layer of sediment that caps the area's flat-topped hills. From the fossils of marine organisms in the limestone, one can see that the sediments now hardened into rock were deposited in the open sea. Below, in the valley, are softer and muddier rocks, deposited earlier when this was a delta close to the seashore. In geological terms, as you walk up from the Dordogne you are first on the delta near the shore, and then you move into the open sea. The succession of the rocks shows that first there was an accumulation of sediments on land and then a rise in sea level.

This much is straightforward geology. All sediments consist of layers of rocks, each with its own characteristic properties. The transitions between the layers represent changes in the environment that led to formation of a particular type of rock—conditions known as the sedimentary environment. The problem is, what kind of changes were they? There are two contradictory views.

Halfway through the last century the French geologist H. Coquand described the entire succession of sediments in southwest France. He could trace most of the transitions between the layers over large distances, and he found very different fossil communities in each of his layers. Coquand believed that the layers themselves represented periods of stability and quiescence, and that the transitions between them bespoke sudden catastrophes of such enormous extent that they wiped out all life on Earth and gave the Creator a chance to start again.

For Coquand, there wasn't one deluge but a whole series of them. His views were representative of a French school of thought, established by the great naturalist and politician Georges Cuvier, which held that the earth's development is marked by a succession of brief periods of intense change and that each such period marks a turning point in history. In between, there are long uneventful periods of stability. Like the French Revolution, after upheaval, everything is different. Likewise, geological time is subdivided into distinct chapters, each with its own basic theme.

A different tradition emerged in the Anglo-Saxon countries. Charles Lyell studied geological forces at work in the present and tried to apply his findings to the interpretation of the old rocks. His adage, "The present is the key to the past," would become one of the cornerstones of geological theory. Lyell was impressed by the immense diversity of the modern earth, including both sedimentary environments and living communities. For virtually every type of sedimentary rock a modern sedimentary environment could be suggested. Here sand was deposited, there clay, and there limestone. Together these environments formed a kind of patchwork, a fluid mosaic. The pattern of the mosaic changed continuously, but the elements remained constant.

According to the Anglo-Saxon school, the boundary between two successive layers did not represent a global catastrophe, but was simply the local record of the shifting transition between two neighboring sedimentary environments. There was no need for global patterns and events. Adherents of this view saw the subdivision of geological time as arbitrary and fundamentally insignificant, like the subdivision of English history into ages named after monarchs.

These views represent the dominant philosophies of the nature of geological history—on the one hand catastrophism, the notion of stability interrupted by brief periods of rapid change, and on the other gradualism, the idea of continuous fluctuation. In Coquand's time, catastrophism was generally accepted in France, but sympathy for this philosophy would soon fade, for purely practical reasons. Geological theory had to be built from scratch. The founders of geology were forced to apply the principle of the present as the key to the past as rigorously as possible. Catastrophism was of little use precisely because it claimed that the geological conditions were fundamentally different from those in the subsequent periods of stability. With the far more advanced geological theory now at our disposal, we can adopt a more flexible attitude. Interestingly, catastrophism is regaining momentum.

Old Gurus and Modern Science

In the late fifties, when I was a student, my companions and I liked to organize debates between wise and experienced scientists about this controversy. The old gurus would fight about the relationships between the hundreds of outcrops that they had studied, but the data were poor, and it was our delight to see them bog down in philosophies that could not be tested. Now, things are different. Immense advances in many areas of earth science enable us to test whether the French were right in saying that the changes they could see in France and elsewhere in Europe were the result of synchronous global, possibly catastrophic events, or were merely parochial fluctuations.

The change came when, due to a combination of factors, attention shifted from the continents to the oceans. First, the developing theory of plate tectonics provided a model for the overall structure and dynamism of the ocean crust (this important concept was amply discussed in the previous chapter). Then the major oil companies started to explore the seas for new energy reserves, bringing with them the resources to develop sweeping new technologies and the determination to elicit cooperation from scientists with widely different viewpoints.

Jan van Hinte, a professor of paleontology at the Free University in Amsterdam, is first of all a geologist whose work for Exxon carried him to Canada, Texas, Libya, and Bordeaux. He owns a charming country castle on the bluffs overlooking the Dordogne, and from there he can freely contemplate the sumptuous landscape that played a part in the great geological controversy of catastrophism versus gradualism. The resolution of this debate was a multifaceted, multibilllion dollar endeavor in which Jan was engaged.

One afternoon, Jan and I settled into hammocks near his stately château. Suspended in the shade of cedars, he explained how the new emphasis on ocean geology has transformed the study of earth history and infused new vigor into the old debate. Compared to the profoundly disturbed geological architecture of the continents, the layered sediments of the ocean floor are relatively undisturbed because there is much less erosion and tectonic disturbance in the sea than on land. He told me stories of three scientific innovations that combined to change our views of the history of the earth.

Innovation 1—Seismic Profiles of the Ocean Floor

As a ship plows through the waves, the drone of its engines is punctuated every 20 seconds by the "Bang!" of a sharp, dry shot emitted by a set of powerful air guns towed behind. Further back, a plastic tube, more than half a mile long and kept at a constant depth of about 30 feet, is dragged through the water. Inside the tube are hundreds of regularly spaced hydrophones, or underwater microphones. The blasts of air, like tiny earthquakes, produce sound waves that penetrate through and are partly reflected by the layers and features of the ocean crust. As these seismic messages are reflected, they are recorded by the hydrophones, relayed to a computer on the ship, integrated, and printed out.

You can listen to the message through earphones: PING ping ping ping, PING ping ping ping. The sound of the air guns echoes back from the transitions between layers of rock in the upper crust. These sounds are incomprehensible, but the computer sorts them out, calculating from each signal the point in the crust from which it was reflected. In this way one track of the boat provides, without drilling, a geological profile—a continuous cross section through the rocks, tens or even hundreds of miles in length, with the position of all the major layers of sediments neatly plotted out. With enough of these seismic profiles you can work out a three-dimensional structure of the subsurface.

"You sail over the ocean all day and see nothing but water," Jan said, "but the printer allows you to look below the ocean floor and detect the sediment structures. Volcanoes, salt domes, faults, subduction zones—they all pass by on the paper scroll. The resolution is about a hundred feet, so you cannot observe anything smaller than that."

Jan fumbled for an enormous sheet [fig. 3.1 (A)] containing one such seismic profile. It was an east-to-west section just outside Cape Hatteras, North Carolina, off the eastern coast of the United States [fig. 3.1 (B)]. He rolled it out on the ground, and I gazed over my hammock at it. Even from a distance it was impressive. In fact, it was better that way. Close up, seismic sections look fuzzy.

The chart revealed the major features of a passive continental margin—one without subduction occurring underneath it. First was the shallow shelf along the coast, the submerged part of the continental mass. Then came the steeper continental slope, inclined towards the deep sea. And, finally, the deep ocean floor itself. When continental

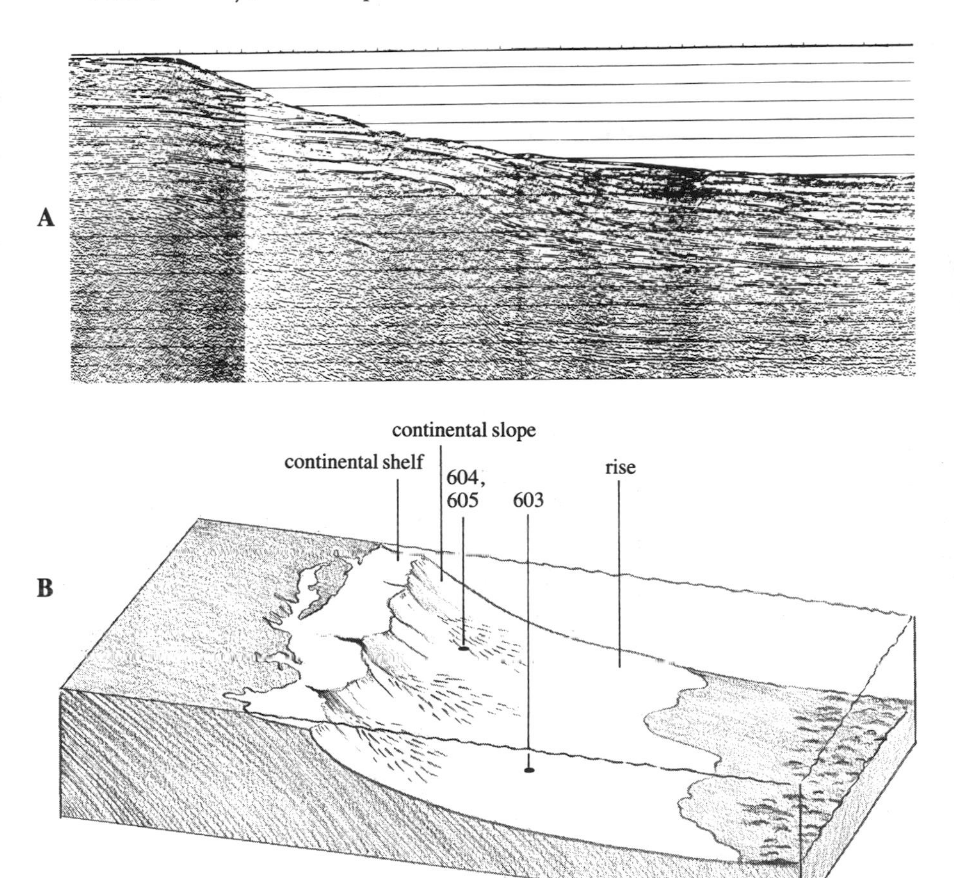

3.1 (A) Seismic reflection profile through the western rim of the Atlantic, near Cape Hatteras, North Carolina. (B) Overview of part of the sea floor, including the shelf, the continental slope, and the rise. Note the position of boreholes 603–605 of Deep Sea Drilling Project Leg 93.

sediments entered the ocean they piled up on the shelf until they were transported over the shelf rim. They coursed down the continental slope and spread out over hundreds of miles of the ocean floor. These aprons of continental detritus at the foot of the continental slope form a gently sloping surface in the deep sea known as the continental rise.

This detailed picture guided a related research effort—an attempt to drill through this portion of the ocean crust. The rise is a key to reconstruction of earth history because the piles of sediment there are a faithful and sensitive recording of all the geological events higher up on the shelf and on land. Every fluctuation in the sea level is beautifully reflected in these sediments deep down on the ocean floor.

Innovation 2—Deep-Sea Drilling

In May 1983, Jan, another paleontologist, Sherwood Wise, and a geophysicist were supposed to head an international team of twenty-one investigators aboard the Deep Sea Drilling Project (DSDP) vessel, the *Glomar Challenger.* The purpose of the cruise was to drill through the continental rise 270 miles east of Cape Hatteras, along the seismic profile we have just seen. Just before they departed the geophysicist became ill, leaving the two paleontologists as co–chief scientists. The absence of a geophysicist was nerve-racking. Such operations cost millions of dollars, and with a seismic profile as the only guide it can be terribly difficult to find the correct location: you set the ship on a calculated course, and use an echo sounder to try to find the right spot. When Jan gave the sign, an acoustic beacon was released and dropped to the sea floor. Fortunately, it hit its mark—there are only three of these beacons aboard, so you cannot make many mistakes. (The beacon serves to keep the ship in the correct position during the drilling.) At 15,000 feet, the drill string reached the sea floor, and finally, drilling could begin.

One after the other, hollow pipes were drilled into the sediment, and then pulled up to the surface. Each piece of pipe retrieved contained a sediment core of the sea floor. The valuable contents were subjected to a preliminary inspection and then carefully stored for more detailed study on land. First, of course, the youngest layers at the surface were sampled; as drilling progressed, older layers were reached. Any moment something could go wrong, and the work would have to stop.

On this DSDP expedition, known as Leg 93, the scientists were lucky. They were able to extract a nearly complete core of sediment ranging in age from the Lower Pleistocene (about 2.5 million years old) down to the Lower Cretaceous (some 130 million years old). The most valuable aspect of deep-sea drilling is that it provides an almost continuous record of geological history. By examining fossils in the cores of rock drilled from the ocean floor, scientists at last can assign ages to the discontinuities (or gaps in the sediment record) between successive layers in the sediment pile. This information is essential if one is to test the opposing ideas on stability and change.

Suddenly, with only 790 feet to go to reach their target, the layer of basalt at the bottom of the ocean floor, the drill string snapped, right on deck. More than three miles of piping worth several million dollars sank to the sea floor. Despite this costly mishap, the crew had gathered enough rock for years of work.

The sediment core from Leg 93 is of special interest because it includes the transition where deep sea and continental realms meet. The immense accumulation of sediments collected from this silent place reflects dramas that took place in the water column and on land. For instance, Jan recalled, "High up in the section we found a massive inflow of continental debris—blocks up to two feet across. The time when those layers were deposited coincides with a severe glaciation in Antarctica. We believe that the glaciers drew so much water from the ocean that the level of the sea dropped, and huge amounts of sediment slid down the continental slope into the deep sea."

Innovation — Refined Dating Techniques

An infinite variety of minute organisms called plankton drifts along in the ocean currents, blooming, feeding, and dying. This life at the ocean surface steadily rains debris that drifts down to the sea floor where it forms a blanket of sediments hundreds of yards thick, growing by an inch or so every thousand years. Due to underwater erosion, there may be gaps in the sediment but nevertheless it is a rich record of geological history, reflecting the climatic, hydrodynamic, chemical, and biological evolution of the oceans. It is this record that can be read layer by layer, inch by inch in the DSDP cores.

Planktonic organisms are widely distributed in the oceans. Thus they allow scientists to construct an ocean-wide standard succession of evolutionary events. Each zone is characterized by specific fossils. With radiometric dating, techniques that measure and make use of the steady decay of radioisotopes, it has been possible to date these fossils and to calibrate zonations with absolute dates—accurate to less than a million years. The sediments usually contain insufficient amounts of these radioactive elements, or isotopes, but by piecing together bits of information it has been possible to date the entire sequence of zones. Now, when we find that a rock contains fossils from a particular zone we can reliably estimate its age.

Some fossils, as seen through a scanning electron microscope, are shown in figure 3.2. The larger shell (it is about 1/100 of an inch across) is a single-celled microorganism called a foraminifer. It lived 105 million years ago. The minute fossils on its surface (fig. 3.2 *right*) are coccoliths: the calcite scales of algal cells. A few grams of deep-sea sediment contain the remains of thousands of foraminifera and billions of coccoliths. These small fossils are ideally suited for detailed statistical analysis. With these refinements in fossil dating techniques, paleontologists can now distinguish between local developments and worldwide geological events.

The fossil skeletons also reveal information about ancient global environments, because their chemistry reflects conditions in the seawater where the organisms were living. Sophisticated instruments measure the ratios of different isotopes of oxygen and carbon in a single shell. Because these ratios are related to the temperature of the water in which the shells are formed, the isotopes allow us to reconstruct the climatic evolution of the geological past.

Hundreds of paleontologists all over the world work to describe and classify the fossil material as it becomes available. When the horizontal distribution of the species are plotted on paleogeographic maps, one can reconstruct the course of ocean currents in the past. But the best thing is that the refinement in fossil dating techniques is a highly

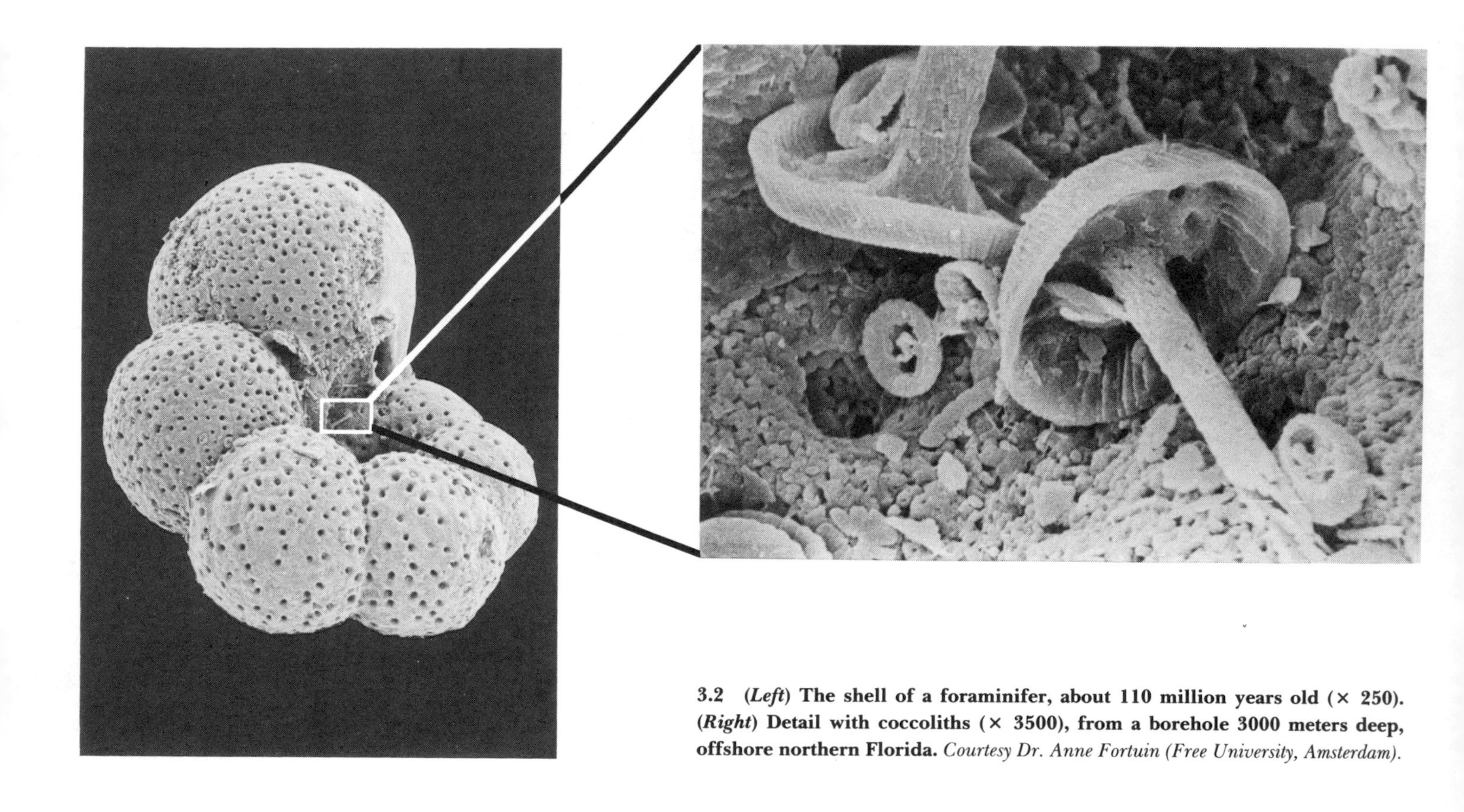

3.2 (*Left*) The shell of a foraminifer, about 110 million years old (× 250). (*Right*) Detail with coccoliths (× 3500), from a borehole 3000 meters deep, offshore northern Florida. *Courtesy Dr. Anne Fortuin (Free University, Amsterdam).*

versatile and sophisticated tool with which to establish the age of the rocks and monitor the history of the earth as a system.

Global Patterns

These powerful innovations—seismic profiling, deep-sea drilling, and improved methods for dating fossils—joined forces in the mid-1960s to change forever our thinking about patterns of earth history. The architect of the new view was a geologist named Peter Vail, who works in Exxon's main laboratory in Houston. While his colleagues concentrated on the thick lines on the seismic profiles (the major marker horizons that highlight the geological structure of most interest to the oil companies), Vail developed a new and controversial approach to interpreting marine seismic profiles. He realized that by looking at the finer details one could extract far more information.

Vail elected to study the irregularities of these linear patterns, places where the traces were broken off and covered up with others with a different orientation. He had considerable sympathy with the French philosophy of global stability and change, and speculated that these breaks are the most significant part of the record—that they represent major geological events which might be of global extent. He gave the breaks names and numbers, so that he could correlate patterns in one part of the world with those elsewhere. And he saw what he hoped for: patterns of sedimentary change that seemed the same all over the earth. Would catastrophism be right after all?

While Vail was incubating a hypothesis that would transform the science of geological interpretation, Jan van Hinte was working at the Exxon research institute in Bordeaux. Jan first became acquainted with Vail's work when Vail visited Bordeaux. "We paleontologists didn't believe a word he was saying," he recalls. "We were all brought up in the Anglo-Saxon tradition of gradual change, and this smelled of catastrophism."

Like all useful hypotheses, however, Vail's could be tested. If his global events were real, they could be dated. If the same horizons everywhere were the same age, synchrony would prove him correct.

Jan had a chance to test Vail's hypothesis himself a few years later. He was in Libya, working on rock samples drilled from the floor of the Mediterranean. When he examined the fossils and the seismic record, he found that the discontinuities in his rocks had the same ages as the global events postulated by Vail. The pattern, Jan concluded, was consistent with Vail's model, and, like a growing number of earth scientists, he was convinced.

The story inferred from the discontinuities is that the level of the planet's oceans fluctuates. Change sea level, and the whole pattern of sedimentary environments changes. Figure 3.3 shows what happens on the shelf, along the continental slope, and in the deep sea at the rise, both during a high and a low stand of the sea. During high stand, the shelf is very wide and acts as the major sediment trap on which the debris coming in from the continents is dumped. Little material will accumulate on the rise; the aprons of sediment will progress very slowly into the deep sea, one in front of the other in what are known as offlap patterns.

When the sea level drops, the load of sediment deposited during the high stand becomes unstable, and starts to move toward the sea. More and more of it is carried over the shelf margin, rushing down the continental slope into the deep sea, where it starts to accumulate on the rise. The figure shows the onlapping patterns of this pile of sediments creeping up the slope.

The team of DSDP Leg 93 wished to drill through the rise because it was there that the most complete sequence

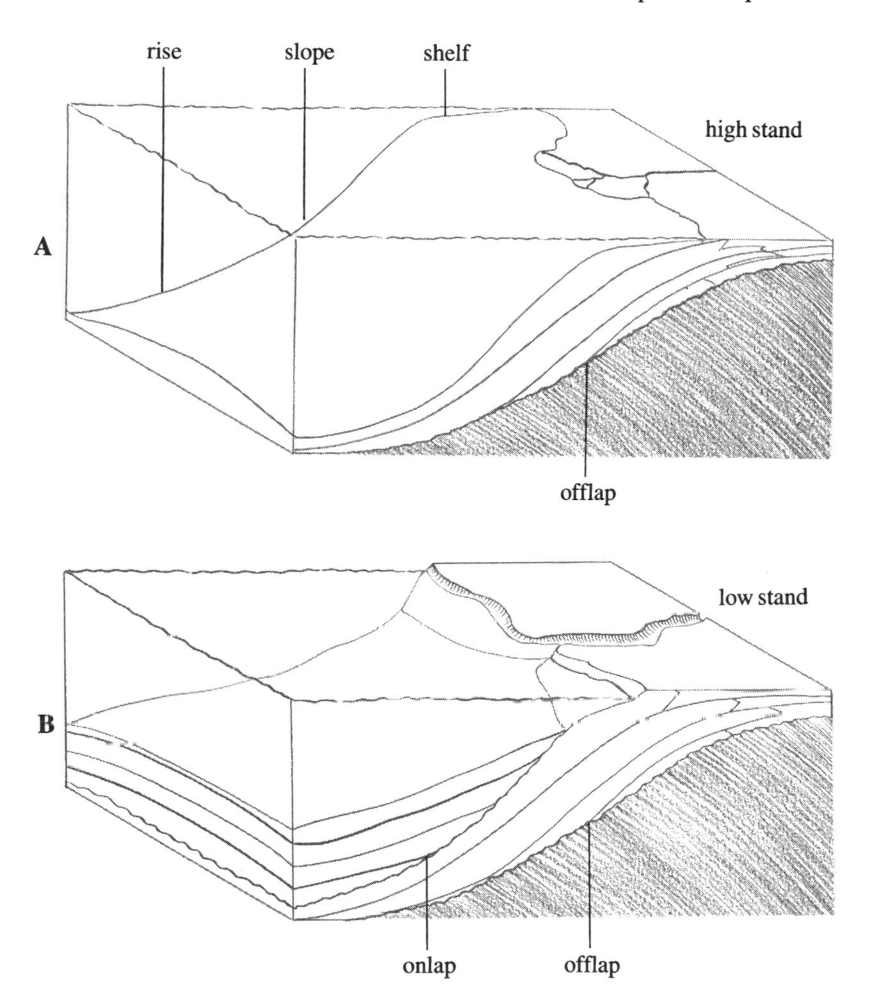

3.3 Principle of the Vail hypothesis: (A) high seawater stand; (B) low seawater stand. *After P. R. Vail et al., American Association of Petroleum Geologists Memoir 26 (1977).*

of offlaps and onlaps, and the best picture of sea-level changes, could be obtained. To the scientists' delight, many of the discontinuities coincided precisely with those on Vail's curve.

The Vail Curve

Vail's contribution was to combine reconstructions of local sea-level curves into a now famous curve of worldwide sea-level changes. Some of the most impressive evidence is here in the lush landscape of the Dordogne Valley. The sequences of rock cycles that Coquand and his colleagues found in France and all over Europe have finally been well dated, using the fossils preserved in the marine sequences. When Vail's succession of sea-level changes is compared with the classic European geological record, the two correlate remarkably well. The succession of rocks here nicely reflects the Vail curve.

So, the French were right all along—some of the breaks in sedimentation really were global events. But the Anglo-Saxon school was also right: The world is a mosaic of diverse environments, and many geological changes can be explained in terms of local or regional, often gradual shifts in the pattern. But it now seems that periods of relatively little change alternate with brief spans of time when the pattern is drastically reorganized.

The world is more complex than the gurus suggested when I was a student—and more interesting. Even Vail was forced to revise his original curves. They showed a saw-toothed pattern (fig. 3.4) suggesting that the sea level comes up gradually and then suddenly falls back again. This puzzled people for many years, until they realized that these are not real sea-level changes, but merely a reflection of sea-level changes in the sedimentary record. Sea-level changes may well be smooth and gradual. Vail has now published a nice wiggly curve. These steady changes have produced the discontinuities in the rocks.

It is a common feature in nature: the drop that makes the bucket overflow. A system that is internally stabilized is gradually undermined by some external influence until it collapses. A small impetus then leads to dramatic change, and an entirely new situation is created. When the sea level is rising, the sediments build up gradually on the continental shelf. When the sea goes down, the sequence becomes destabilized. It hangs on for some time, and then—Wham! Part of it slides into the deep sea. Eventually, sea level again begins to rise and bit by bit, the sediment builds up.

The Nature of Geological Change

The simplest view of earth history that may be derived from the Vail curve is of an endless repetition with no real development: gradual sea-level changes causing sudden fluctuations in the distribution of sedimentary environments.

The real situation is probably far more complex. The changing sea level is thought to have dragged along many other changes in its wiggling course. What exactly these changes are is a matter of intense speculation at present. Times of high sea level are thought to be correlated with warm climates and low sea levels with cool ones. High stands tend to cause stagnant oceans, depleted of oxygen, whereas at low stands the water is believed to be circulating and well aerated. At high stands there is little erosion, and as a result the ocean waters are poor in nutrients. In contrast, at low stands there is large-scale erosion and large amounts of

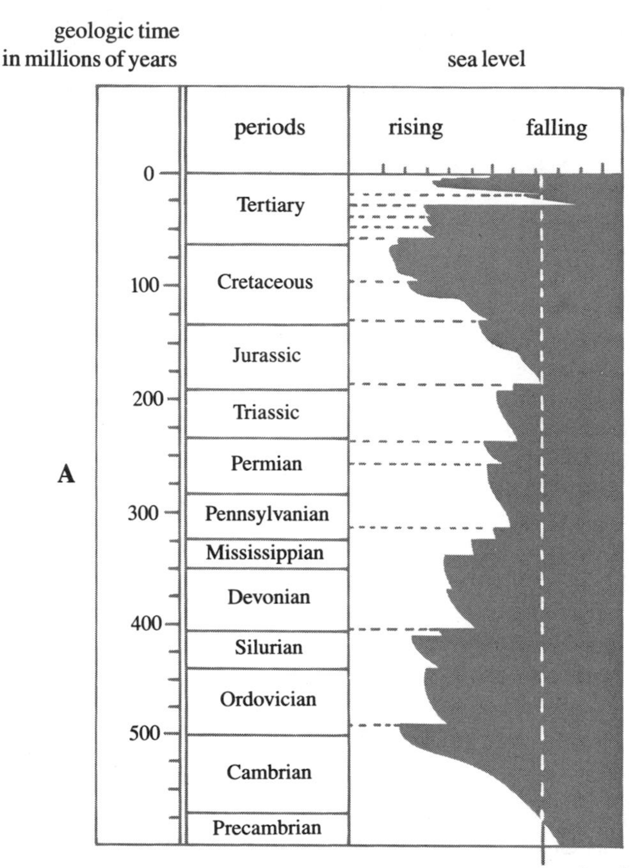

3.4 (A) The Vail curve: a generalized curve of the last 570 million years. *Adapted from Bilal U. Hag, Chronology of fluctuating sea levels since the triannic,* Science *(235:1156–1175; 1987). Copyright 1987 by the AAAS.*

nutrients enter the ocean. Such changes must have had a major impact on the evolution of life.

I can fully agree with Jan van Hinte that the ongoing beat of sea-level change has been a major forcing function for evolution, alternately constraining and boosting the proliferation of life on earth. Yet this view suffers from the same one-sidedness that we encountered in our discussion

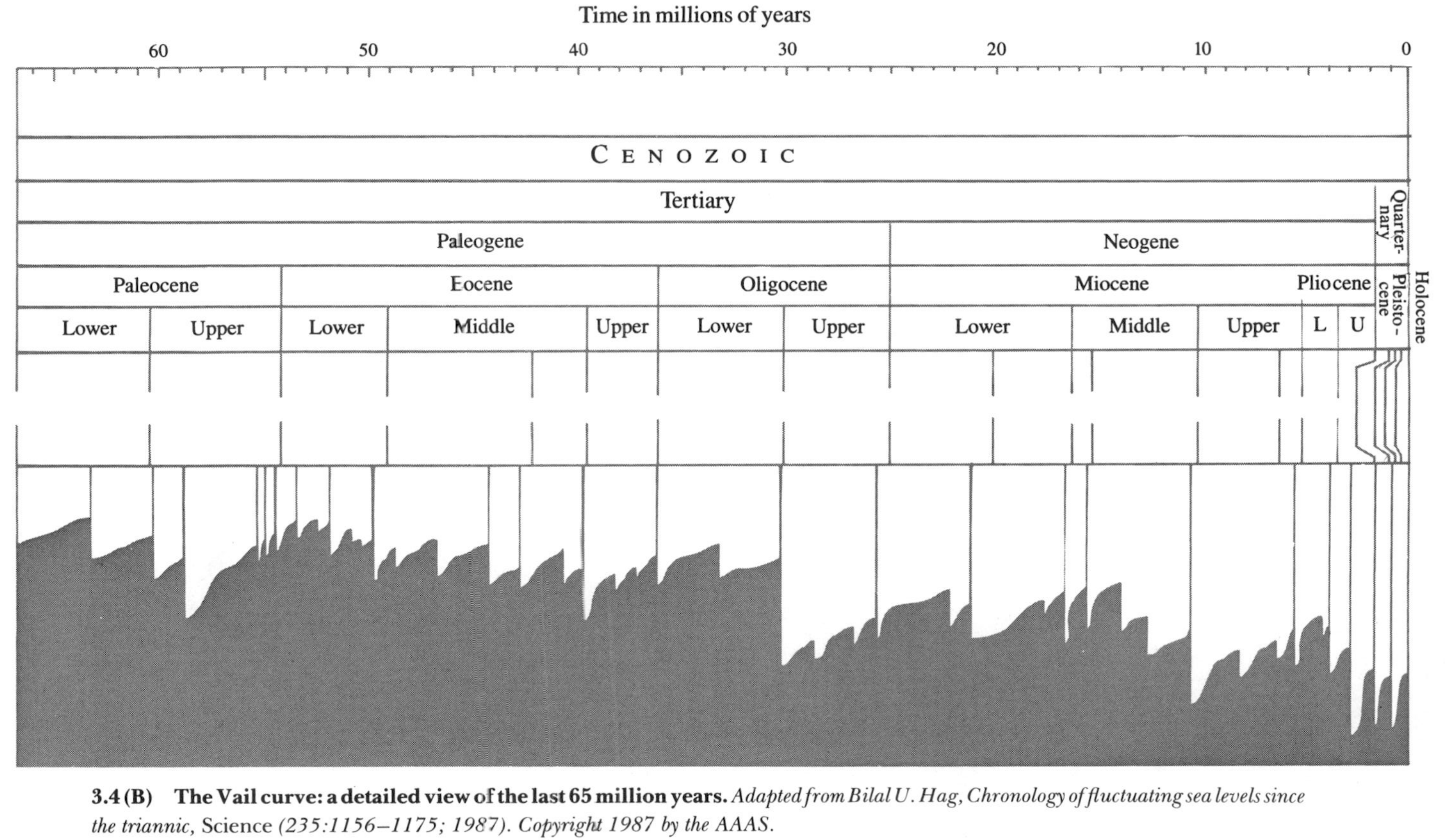

3.4 (B) The Vail curve: a detailed view of the last 65 million years. *Adapted from Bilal U. Hag, Chronology of fluctuating sea levels since the triannic,* Science *(235:1156–1175; 1987). Copyright 1987 by the AAAS.*

of Scotland. The biota—all plants, animals, and microorganisms taken together—is envisaged as merely adapting to given physical and chemical conditions, while the active intervention of life itself is ignored. Hence, life forms preserved in the deep-sea sediments such as the coccoliths and foraminifera shown in fig. 3.2 have been treated as mere tools for reconstructing the geological past.

Interestingly, the algae that build the elegant coccoliths on which geologists rely are among the most telling examples of the huge forces of life. In chapter 6 I shall discuss this curious case in some detail. (Recently, and to my utmost delight, Jan van Hinte and I have agreed to study together the role played by these organisms in shaping the global environment.)

4

THE STYLIZED WORLD OF ROBERT GARRELS

IN 1971, geologist Robert Garrels infused his science with a large dose of chemistry. He published a book that transformed the way geologists look at our planet. In keeping with the new thrust of those days—the theory of plate tectonics—*Evolution of Sedimentary Rocks* took a global view of the earth, but was unusual in that it treated the earth as a huge chemical factory. Through a series of lucid arguments, Garrels and his coauthor Fred Mackenzie distinguished between earth's major chemical reservoirs and the transport pipes and conveyor belts that link them, and considered the bewildering cycling of matter through this complex maze. Not only did they outline the production system, but they also approached the cycling problem from a quantitative point of view.

When the earth is viewed this way, all kinds of questions crop up. For instance, how much limestone is there on earth, how much iron, sandstone, gypsum? What is the mass of all the sediments? How much sodium is liberated each year by weathering, and transported by the rivers into the oceans? And how long does it stay there before it is removed again? These questions are significant because the earth is a system closed to matter: there *must* be finite amounts of all its constituents. Answers to these questions are essential if we are to understand our planet.

Collecting this kind of information is another matter. For the atmosphere and the oceans, it is relatively easy. Their masses can be estimated and the concentrations of their ingredients measured. But how about the heterogeneous hodgepodge that the lithosphere happens to be? No one could ever collect a representative sample of rock and weigh, for instance, the amount of sulfur in it. All one can do is to make an estimate, based on reasonable assumptions and a limited number of measurements. This is the basic weakness underlying global cycling research. You often have to work with crude approximations and hope for the best.

Garrels owed some of his encyclopedic knowledge of geology to this weakness. It forced him to search relentlessly for more reliable information and criteria. Also, when you consider the global dimensions and complexity of this factory, and add to that the overwhelming periods of time over which the factory has operated, you begin to understand why Garrels had to develop a remarkable power of simplification. His study of the coupling between the earth's chemical reservoirs was, after all, only one aspect of his academic research. His mind must have been highly selective. How did he tell the important from the unimportant, and make complex things simple?

The painter Piet Mondriaan once said that the art of art is to simplify, to bring to the fore the essence of what we see and leave out the subordinate. And indeed, over the years Mondriaan's paintings became less and less elaborate. They faded away until only a few square surfaces remained. So it was with Robert Garrels. With his erudite arguments, he silenced the squabbling of generations of geologists and brought into the open the major problems needing consideration.

Robert Garrels died in 1988. By the end of his life, he had launched the science of geochemistry of the outer earth, and had given a global perspective to this field. Indeed, through his scientific models and other contributions we can glimpse a fundamental problem: How do the geological forces of life fit into global and historic perspectives?

Figure 4.1, modified from an important paper published in 1974 by Garrels and Ed Perry, represents the global geochemical factory. There are twelve spheres, some big and others very small. They represent the global reservoirs of the predominant constituents of the outer earth. These reservoirs are entangled in a maze of tubes that represent the fluxes of matter between them. The sizes of the reservoirs and fluxes are known, at least approximately. Note that two very small reservoirs occupy a central position and are linked with all the others. They are carbon dioxide (CO_2) and free oxygen (O_2). These substances are located in the atmosphere or occur in a dissolved form in the oceans, while the other reservoirs are in the lithosphere.

In order to appreciate Garrels's view of the earth, it will help to develop a general feeling for the kinds of substances in the spheres. Those marked $CaSiO_3$, $MgSiO_3$ and $FeSiO_3$ refer to the silicate minerals contained in rocks emanating from the deep earth: basalts, granites, and the like. Although they are of utmost importance in earth dynamics and play a large part in Garrels's considerations, they are of little consequence for the present argument. SiO_2 is best known as the glass-like mineral quartz, which also is common in granites, but in the oceans it constitutes the delicate

4.1 Garrels's and Perry's stylized view of the global geochemical factory: chemical reservoirs with fluxes in between.

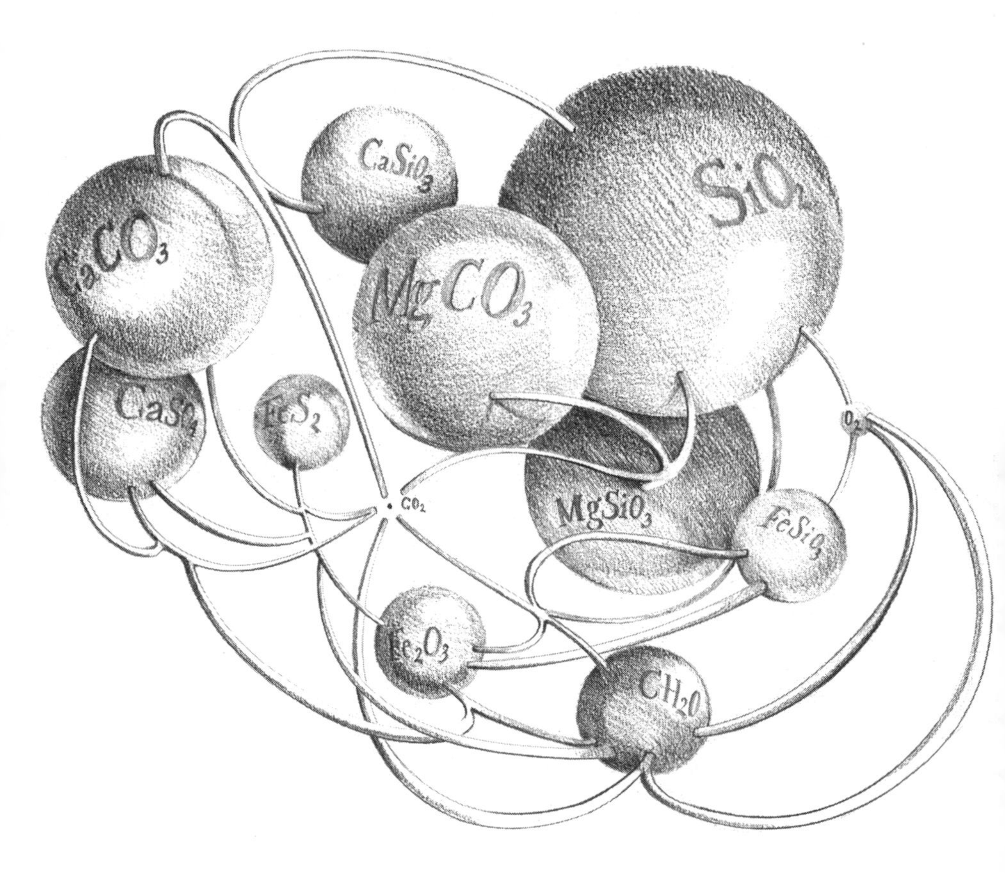

skeletons of diatom algae. Calcium carbonate ($CaCO_3$) and magnesium carbonate ($MgCO_3$) together stand for limestone. The huge underground reservoir of organic carbon, petroleum, and coal is symbolized by CH_2O. When this material reacts with oxygen, it is converted into water and carbon dioxide, and the latter of these, in turn, can combine with calcium or magnesium to form limestone.

One constituent, iron sulfide—FeS_2, or pyrite—may be less familiar. In sedimentary rocks it usually occurs together with organic carbon, and may give the rock a distinct black or gray cast. Large crystals of pyrite have a beautiful golden luster, and adorn many amateur rock collections. Pyrite, a by-product of decomposition of organic remains in the sea, derives from bacteria. When exposed to the air and some water, it readily decomposes. The iron turns into rust (Fe_2O_3) and the sulfur into sulfate, a common constituent of seawater. Finally, when seawater evaporates, in salt ponds for instance, dissolved sulfate and calcium will combine to form the mineral gypsum ($CaSO_4$).

These are the major reservoirs. Now imagine the whole system at work. The constituents in all the reservoirs react with one another. The compounds that are formed flow towards the appropriate reservoir. Carbon is channeled back and forth between the lithospheric reservoirs of limestone and organic carbon, and all this traffic passes through the tiny reservoir of carbon dioxide in the atmosphere and the oceans. Oxygen is liberated when organic carbon and pyrite are buried, and so on. Materials stream continuously through the system. At first sight, it evokes the maze of retorts and tubes typically used in comic papers to portray the mad professor's laboratory.

Turmoil

Michael Whitfield, of the Plymouth Marine Laboratory, England, had the idea of condensing the entire lifetime of the earth, 4.6 billion years, into a single 24-hour day running from midnight to midnight. On this contracted time scale, accretion of the earth started at midnight and ended shortly after 4 AM, when the oldest rocks we know were formed. By 6 AM there was life. Appreciable amounts of free oxygen built up in the atmosphere by about 11 AM. Eukaryotic cells made their entry at about 5 PM; the first animals appeared at 8:20 in the evening; and the Phanerozoic, the period of earth history when well-organized calcified fossils appeared on the scene, began at 9:10. Vascular plants colonized the continents at 9:50. At 10 minutes to midnight, humans arrived, and the industrial revolution took place at 3.7 milliseconds before midnight.

On this time scale, global cycling makes a memorable impression. Table 4.1 includes the two atmospheric reservoirs, oxygen and carbon dioxide, as well as some important components of the oceans and the lithosphere. The lithospheric reservoirs are measured in hundreds or even thousands of units, while the oceanic and atmospheric reservoirs are much smaller (76.5 units or less), and the amount of carbon dioxide in the atmosphere is next to nothing (0.05 units).

The middle column of table 4.1 shows the time that the global system needs to recycle each of its constituents. The ponderous lithospheric reservoirs are sluggish, yet on our contracted time scale these huge masses are renewed in a few hours. (This does not mean, of course, that everything is recycled and that nothing is left of the older stuff. A lot of it remains—to the delight of geologists—because some

Table 4.1 Dynamism of the Earth
(seen on a highly contracted time scale)*

	Reservoir size (x 10^{18} mol)	*Time needed to recycle present mass if earth were 1 day old*	*Times recycled during last 600 million years*
		ATMOSPHERE	
biological cycle			
CO_2	0.055	0.2 millisecond	55,000,000
O_2	38	140 millisecond	80,000
geological cycle			
CO_2	0.055	37 millisecond	300,000
O_2	38	2.2 minutes	76
		OCEAN	
bicarbonate	2.8	1.5 seconds	7700
sulfate	40	8.2 minutes	23
calcium	14	13 seconds	857
magnesium	76.5	4.5 minutes	41
		LITHOSPHERE	
ocean floor	—	30 minutes	6
total sediment	—	2 hours	1.4
CH_2O	1250	1 hour, 40 minutes	1.8
FeS_2	250	5 hours	0.6
$CaSO_4$	250	1 hour, 20 minutes	2.4

*The entire period that the earth has been in existence (4.6 billion years) is taken to have lasted only a single day. The sizes of a number of atmospheric, oceanic, and lithospheric reservoirs are given in 10^{18} mol—a measure of the number of molecules they contain. On the contracted time scale, the time needed to recycle the present masses of the global reservoirs varies between a fraction of a millisecond and several hours. In the right-hand column, the number of times that each of the reservoirs has been recycled during the last 600 million years is indicated.

portions are used up more than once during the recycling period.) In comparison, the outer earth is a hissing boiling pot, total pandemonium. The oceanic and atmospheric constituents are recycled in a matter of minutes, or even seconds. The chief agitators are carbon dioxide and oxygen. Carbon dioxide, of course, is used in photosynthesis and released again by respiration, and the situation is reversed for oxygen. In the biological cycle, oxygen is renewed every seventh of a second, and carbon dioxide five times every thousandth of a second. In other words, the biological cycle of oxygen turns at the speed of a simple table centrifuge—420 revolutions per minute, while the carbon dioxide cycle turns with the speed of an ultracentrifuge—30,000 revolutions per minute.

Robert Garrels ignored this biological cycle in his models, reasoning that its immediate effect on the large-scale geochemistry of the earth is negligible. Oxygen and carbon dioxide come and go and that's that. But 0.1 percent of the organic carbon produced by photosynthesis is buried with the sediments. In our contracted earth history, it remains inside the gigantic reservoir for 1 hour and 43 minutes before it reaches the earth's surface and is reconverted to carbon dioxide. Oxygen is returned to the atmosphere by the same process and then removed again. In addition, carbon dioxide is drawn into the lithosphere by weathering reactions, and returned to the atmosphere by volcanism and metamorphism (transformation of rocks in the earth's interior by high pressures and temperatures). Although the pace of this geological cycling is much slower than that of photosynthesis and respiration, it is still extremely rapid. Carbon dioxide is replenished every 25th of a second, and oxygen every 2.2 minutes.

Are the Reservoirs Rigid, or Do They "Breathe"?

If figure 4.1 represented a system of rubber balloons connected by plastic tubes, what turmoil would be occurring inside! It is difficult to conceive how the whole construction could *fail* to become distorted after a while. The geological record suggests, however, that the terrestrial system hasn't changed dramatically over hundreds of millions, or even billions, of years. The outstanding argument in favor of a stable earth is that (on our contracted time scale) it has supported life without interruption from 6 AM onwards, and animal life for the past 3½ hours. This severely limits the variations allowed in the global environment. Garrels and Perry could not conceive that concentrations of atmospheric carbon dioxide and oxygen had varied very much, particularly in relatively recent times. Large variations would have resulted in such adverse conditions that animal life would have been destroyed.

This is a pervasive problem of global cycling, the contradiction between frantic movement occurring inside the terrestrial factory and the almost perfect quiescence of its workings. Over the years, this problem churned through Robert Garrels's brain; he began to liken the earth system to a grandfather clock, in which the rotation of one wheel determines the turning of all the others.

There are two possible modes of operation. One is that the balloons and their connecting pipes are not as malleable as rubber but as rigid as steel. In that case, the system would not change over time: the same amount of material goes into each reservoir as is allowed to go out. In the second, the balloons and fluxes change over time, but the movement is such that the biosphere would not be severely affected.

The geochemists have found a few reliable clues by which they can decide which of the two alternatives is correct. Figure 4.1 shows that there are two major reservoirs of sulfur on earth. One is gypsum ($CaSO_4$), the other is pyrite (FeS_2). Sulfur is sulfur, you may think, but with sophisticated apparatus the geochemists can distinguish between different kinds. They measured the ratios between these different isotopes in fossil gypsum layers of many different ages and found that they tell about the relative sizes of these reservoirs in the geological past.

Figure 4.2 *left* plots these ratios through time, a curve with sharp peaks and deep troughs. Where the curve moves to the right, the lithospheric pyrite reservoir grew at the expense of the gypsum reservoir, and where it goes to the left, the pyrite reservoir shrunk and the gypsum reservoir swelled. And since the sum of the sulfur in the two reservoirs is assumed to be constant throughout geological time, their movement must have been perfectly coordinated. It is like two balloons with a pipe in between. You compress one, and the other one expands.

So, the reservoirs of pyrite and gypsum *did* change. They made a kind of "breathing" movement during the past 600 million years. Toward the end of the Precambrian there was a huge shift from sulfate (gypsum) to sulfide (pyrite). Then there was a gradual increase in sulfate again until a very high peak was reached in the Permian, about 220 million years ago, and then the system shifted back to sulfide. Superimposed on that general movement were all kinds of hiccups, big and small. The reason I compare this movement to breathing is that, on the whole, it did not go in any particular direction. There was no trend towards more sulfate or sulfide, just a movement to and fro. (Because the contents of each reservoir could be recycled several times

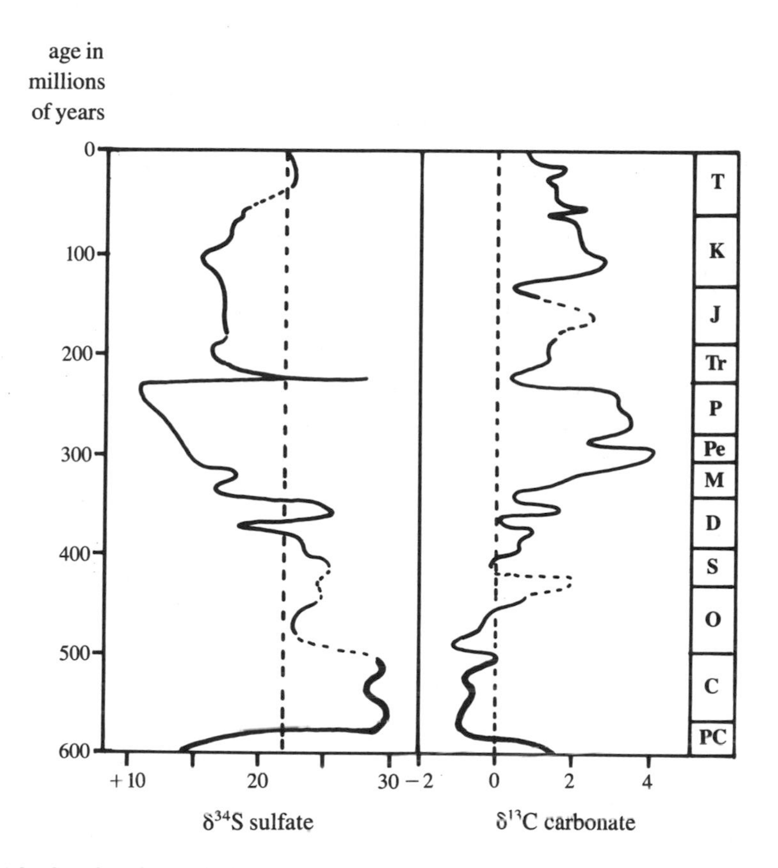

4.2 Secular change in isotope composition of sulfur (*left*) and carbon (*right*). A trend to the left in the left-hand curve represents a net transfer of sulfur from the pyrite (FeS_2) to the sulfate ($CaSO_4$) reservoir (compare fig. 4.1). This reaction sucks oxygen away from the atmosphere. A trend to the right in the right-hand curve represents a transfer of carbon from the limestone ($CaCO_3$) to the organic carbon (CH_2O) reservoir. This reaction pumps oxygen into the atmosphere. Note that, on the whole, the consumption of oxygen by one reaction is offset by a liberation of this gas by the other, so that the oxygen content of the atmosphere remains approximately constant through time. For an explanation of the abbreviations on the geological time scale see figure 3.4 (A).

during one interval, however, in some respects breathing is an inadequate analogy.)

Garrels and Perry noticed another implication of the isotope curve: oxygen would be used in shifting sulfur from the sulfide to the sulfate reservoir. They made a quick calculation and came to the startling conclusion that the transfer which took place between the Cambrian and the Permian might have cost *several times* the amount of oxygen present in today's atmosphere. Nevertheless, throughout that period animal life developed without any major disturbance, despite the enormous drain on the oxygen reservoir, so the atmosphere must have remained more or less constant. The problem was formidable. If the lithosphere was "breathing" at that rate, how could the atmosphere remain unchanged? What mechanisms kept it in check?

Garrels and Perry realized that to solve this major question was asking too much at that time. Then they had a very fruitful idea. If they reversed the problem—put it upside down as it were—a question emerged that could be tackled straight away. Let us accept the fact that the composition of the atmosphere was constant, they reasoned: What would be the effect of this constraint on the "breathing" of the lithosphere? They came up with a golden rule that has dominated the global cycling debate up to the present. If oxygen were sucked away from the atmosphere by any net transfer of sulfur from the pyrite to the gypsum reservoir, then it had to be returned through the lithosphere. There was only one possibility: oxygen was liberated into the air when limestone was changed into organic carbon, and this had to compensate for the oxygen lost during the shift between reservoirs.

The significance of this is startling. If Garrels and Perry were correct, then not only have the sulfide and sulfate reservoirs been "breathing," but so have the reservoirs of limestone and organic carbon—and in perfect coordination. According to their formula, for each 4 molecules of pyrite changed into gypsum, exactly 15 molecules of limestone (8 of $CaCO_3$ and 7 of $MgCO_3$) have to be removed from the lithosphere, and precisely 15 molecules of organic carbon are stored in it.

This does not imply that the atmosphere and the oceans have not been playing a part in the "breathing" of the lithosphere. On the contrary, they were actively involved. When sulfur is transferred from the pyrite to the gypsum reservoir, oxygen is sucked out of the air. All that is said is that an equal amount of oxygen is brought back to the atmosphere by the net transformation of carbonate to organic carbon. In other words, the atmosphere and the oceans act as a medium through which the oxygen is distributed between the different reservoirs, but their oxygen content remains unchanged.

The same is true for carbon dioxide. This gas is injected into the atmosphere when limestone is formed in the oceans, and it is sucked out again by photosynthesis and the subsequent storage of organic carbon in the sedimentary rocks. The two reactions are exactly balanced so that the concentration of the gas remains unaltered. Garrels and Perry proposed that a tight coupling of the sulfur and carbon cycles is needed to maintain a constant ocean-atmosphere system.

Here, then, is the crux of the argument for chemical cycling on our planet: Garrels and Perry believed that, in spite of all the fluxes in the global system, the reservoirs are stable and all changes take place in a coordinated fashion. From the sulfur isotope curve, they concluded that the pyrite and gypsum reservoirs are engaged in a breath-like movement. And finally, the requirement of a constant atmosphere and ocean, deduced from the fossil record, forced them to assume that four different lithospheric reservoirs had been "breathing," in perfect coordination. Further analysis revealed that eight lithospheric reservoirs were actually involved.

Coupling of the Sulfur and Carbon Cycles

What a weird description this is of geological history, and how far removed from the ordinary practice of the geologist in the field. It is a surprising picture, assembled from a few well-chosen data. But are we in danger of losing ourselves in abstract arguments? Is there any independent proof to show that this highly unlikely scenario has any relation to the real world?

Again, the answers come from isotopes, this time in the carbon of limestones. They tell about the relative sizes of the reservoirs of limestone ($CaCO_3$ and $MgCO_3$) and organic carbon (CH_2O) in the geological past. Look at figure 4.2 *right* which portrays these trends through time. Bear in mind that a trend to the right in the curve indicates a net transfer of carbon from the limestone to the organic carbon reservoir, and a trend to the left indicates a transfer in the opposite direction. This curve describes the antithetic "breathing" of these two reservoirs.

At the time that Garrels and Perry wrote their paper this latter curve was not yet available, but by the early 1980s, the data were in and a new chapter in the modeling saga could begin. The question was, Do the two isotope curves show that the cycles of sulfur and carbon had been coupled throughout the past 600 million years?

When you lay the two curves side by side, as in figure

4.2, you immediately see that, on the whole, the sulfur and carbon isotope curves mirror one another, just as the formula requires. Such a simple observation is not very satisfactory, however. One would like to derive *quantitative* estimates of the reservoir transfers from the curves, and see whether they agree with the model. This was the next goal that Garrels and his associates set themselves.

The problems were great. Somehow, the evolution through time of the four lithospheric reservoirs under consideration had to be derived from the isotopic curves. The old representation of the global factory was so complicated that it didn't help. A much simpler system was needed. Garrels teamed up with his friend Abe Lerman, and together they worked out a model that described only the global carbon-sulfur cycle and omitted all the others.

Figure 4.3 gives the outline. In the center of the figure are the ocean and the atmosphere. In the ocean, two reservoirs are important, one of sulfate and the other of dissolved limestone, here represented as carbonate. In accordance with the basic assumption of coordinated "breathing" they are kept constant throughout the calculations. The lithospheric sulfur cycle to the left shows the reservoirs of gypsum and pyrite, and the fluxes from the reservoirs to the ocean-atmosphere system and back. In a similar way, the figure of the carbon cycle, on the right, relates the reservoirs of carbonate and organic carbon (CH_2O) to the oceanic dissolved limestone.

The figure sums up the situation on earth today. Using some straightforward geochemical assumptions, the model could be hooked up with the sulfur isotope curve of figure 4.2 *left*. In fact, the curve served as a kind of rail along which the whole system could be shifted back in time from the present. It was an amazing experiment, a computer simulation of the history of our planet over 600 million years. At any moment throughout this time span the whole system was determined. One could see the four lithospheric reservoirs "breathing," following the ups and downs of the sulfur isotope curve. Going from modern times towards the Permian, about 250 million years ago, the reservoirs of gypsum and organic carbon grew to gigantic proportions, whereas the pyrite and carbonate reservoirs shrunk proportionately. Then the trend was reversed until the end of the Precambrian, about 600 million years ago, when the system again was forced in the opposite direction.

Garrels and Lerman wanted to derive a purely hypo-

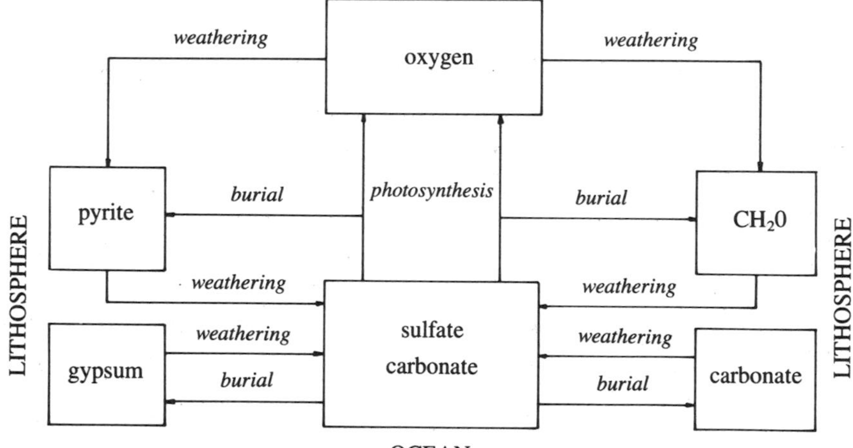

4.3 Model for coupled carbon and sulfur cycling. The ocean-atmosphere system in the center is kept constant. The lithospheric sulfur cycle on the left is hooked up with the sulfur isotope curve of figure 4.2 *left*. A hypothetical carbon isotope curve is calculated by the computer and compared with the real carbon isotope curve of figure 4.2 *right*. *Adapted from R. M. Garrels and A. Lerman, Coupling of sedimentary sulfur and carbon cycles—an improved model,* Amer. Jour. Sci. *(284:989–1007; 1984). Reprinted by permission of American Journal of Science.*

thetical carbon isotope curve from the actual sulfur isotope curve, under the rigid specifications of a constant ocean and atmosphere. They reasoned that if this hypothetical curve corresponded with the real one, then the simplifying assumptions underlying the whole operation could not be too far off beam.

Their only dependable clues were buried in the isotopic data for sulfur and carbon. When they used the sulfur isotopes to predict the carbon data, to their delight there was reasonable agreement between the actual values of the carbon and what they had calculated. There were differences, Garrels recalled,

> . . . but it was 1980, and in those days the carbon data were not very good. Still, we became reasonably confident that the carbon and sulfur cycles were coupled, and that, with all our assumptions, we had been on the right track. Then a disturbing thing happened.
>
> Bob Berner had run the same system, but the other way around. He had started with the Cambrian, 600 million years ago, and moved from there to the present. Of course, he hadn't a clue what conditions in the Cambrian were like, but he reasoned that if he repeated his run often enough, changing the initial conditions each time, he would eventually end up with the real world of today. That last run would then represent a reconstruction of the entire history, but running from the past to the present, and not the other way round, as we had tried. The problem was that Bob Berner came up with different conditions for the Cambrian than those we had predicted.

For two years Garrels and Lerman were at a complete loss.

> Then, Abe came down for a week, and it was almost like Kekulé, who invented the carbon structure of the benzene ring in a dream. . . . In *my* dream I was running motion pictures backwards. A guy, upside down, shot out of the pool and landed on

the diving board. Water splashed up from all sides and filled the hole left by the diver. And I understood: To run the earth backward, you must *really* run it backward and not forward. Rivers run out of the sea to their sources. Rain falls up into the clouds. Salt dissolves from the saltpans into the ocean. It is sucked up by the river and brought up to the mountains. And there it disappears into the rocks.

This dream provided the essential clue. All Garrels and Lerman had to do was reverse the signs of all the mathematical formulations for the fluxes, change the positives into negatives, and vice versa. When they reran their computer model, the results showed beyond any doubt that the carbon and sulfur cycles are coupled, just as Garrels and Perry had assumed years before. Granted, the coupling was not that tight; there were many minor discrepancies, but these could be explained on the grounds that the two cycles operate at different rates.

How the Earth Works

So, at long last Garrels and Lerman had settled the problem. The atmospheric oxygen levels are kept in check because the sulfur and carbon cycles are intimately coupled. Over the last 600 million years the reservoirs of organic carbon and limestone, and those of pyrite and gypsum, have been "breathing" in an almost perfectly antithetic fashion, so that stable oxygen levels were maintained in the atmosphere. But the question remains: How could the system be kept in check for such a long time?

Garrels initially felt that the operation of his system was essentially automatic, only held in check by a fortuitous constellation of haphazard stabilizing mechanisms. He realized that life influences the earth's dynamics in many important ways. For instance, without photosynthesis the reservoirs of organic carbon and oxygen would not exist. In addition, living organisms bring about a smooth flow of most of the fluxes in his model earth. But it seemed to him that control of the gigantic machinery of the earth was a feat even beyond life itself. The biosphere had to adapt to the conditions imposed by the system, and the stability of ocean and atmosphere was a matter of good luck.

Or does the structure of pipes and balloons that Garrels took to represent earth's dynamics conceal a highly organized entity, one that emerged from the inanimate primitive earth and then propagated itself through geological time? Is there a regulatory system that doesn't show up in the models?

Near the end of his life Garrels began to suspect that this might be the case. Although his models start from the assumption that the earth's dynamics result largely from the haphazard interplay of physical and chemical forces, he was open to the idea of a more organized world.

PART II

Life, the Missing Link

5

LIFE AND THE CYCLING OF ROCKS

GLOBAL GEOLOGY allows us to imagine the earth from a lookout in space with time condensed to the point where millions of years elapse in mere minutes. The continents move majestically over the globe like sailboats gliding over a lake on a warm summer afternoon, each following its own circuitous path. Plates shift and turn around; at their margins, ocean floors are created and destroyed while volcanoes erupt. News archipelagoes emerge from the deep and are swept along enormous faults towards the continental masses. In one instant, the continents scatter apart. A while later, they all seem to return along their old pathways and reassemble into a single, massive block. But even this situation does not persist. The supercontinent breaks up again, the fragments disperse, and the cycle begins anew.

Two continental masses approach each other, their collision inevitable. On a compressed time scale, the rocks appear to be plastic. They contort and squeeze up like pastes from the deep earth. Cracking and grinding, a huge mountain range emerges along the entire length of the collision zone.

As soon as the uplifted rocks appear at the surface, they begin to disintegrate, fiercely attacked by water and air. Even the hardest of rocks is weathered away. Thick aprons of debris and dust wash off the mountains and glide down the slopes into the ocean. Meanwhile, fresh rocks well up from the deep, replacing their decayed predecessors and replenishing the fluxes of debris. Steadily, older and deeper recesses of the crust are uncovered. It is as if the planet were turning itself inside out. The emerging mountains are involved in a constant process of renewal and erosion.

The debris does not flow down the mountain slopes in a steady, continuous stream. Rather, it moves in a series of collapses controlled by the beat of changes in the sea level. And, once under water, the debris flows from the continental shelves down to the deep sea. Even there it does not come to rest but is swept along by the steady shift of the plate-tectonic conveyor as new crust is carried away from the mid-oceanic ridges. This conveyor pushes the debris into deep-sea troughs where it is scraped off or carried into the deep earth to be transformed into molten rocks. This material is pushed up, and enters a new phase of spasmodic mountain formation.

This is the way of the world. Rocks are involved in a cyclic process of renewal and disintegration. Their melting and subsequent solidification is followed by uplift, exposure, weathering, erosion, transport, sedimentation, and transformation, until a new phase of melting sets in. The concept of rock cycling was first advanced more than two hundred years ago by James Hutton, one of the founders of modern geology. Plate tectonics adds a global dimension to this concept, and describes the basic mechanisms. Geochemists such as Robert Garrels emphasize the deep involvement of the atmosphere and hydrosphere, and they model the cycling of rocks, water, and air as a coherent interactive process.

Do these cycling models adequately describe the earth's dynamism? How deeply is the process influenced by the multiple catalytic effects of evolving life? The cycling of rocks must now be regarded from the biological point of view.

The Cycle in Motion

The complexities and many facets of the rock cycle are in abundant evidence on Vlieland, the second in a string of islands stretching from the northern tip of Holland along the coast to Denmark (fig. 1.2). These islands form an elegant barrier that prevents the mud flats and coast behind from being destroyed by waves that would otherwise sweep in from the North Sea. As the barrier continues south along the Dutch and Belgian coasts it becomes a virtually unbroken zone of dunes, only a few miles wide, that protects the lowlands from the sea: a natural wall, of vital importance for the integrity of both countries, and carefully maintained over the centuries.

The dunes are a curious mixture of wild nature and

artifact. Frank van der Meulen, a biologist and geomorphologist charged with the management of the dune landscape near The Hague, explains that the present system of dunes began to form in the twelfth century. Huge amounts of sand and mud had been mobilized by the large-scale erosion of forest soils following intensive tree-cutting in the European hinterland. The debris flowed to the sea by means of the Rhine and the Meuse. There it augmented the sands left from the last ice age, and was swept north and east by the sea currents that move along these coasts. Thus, a human-induced flux of ancient soils appears to have provided much of the raw material for the present dune barrier.

Although they protected the inhabitants of the lowlands behind, the dunes have been an unruly friend. Entire villages were buried in the untamed fluxes of sand. Agricultural ventures were unsuccessful. The few crops that could grow on the nutrient-poor soils were covered with sand or devoured by rabbits. The dune belt has remained an uninhabited strip of wild land.

Wild, yet carefully held in check. At all costs, a breakthrough of the sea had to be prevented and the damage caused by moving sands had to be restrained. Legislation dating from the fourteenth century and still in force rules that no sand should be allowed to be tossed around by the wind. Wherever the sandy substrate was exposed by storms and rabbits, reeds or marram grass had to be planted to cover it. The native biological forces, specially equipped as they were to facilitate dune formation, were mobilized to fix the landscape for eternity—or even to extend it. For instance, the island of Texel, the first in the string, was manufactured by joining two islands together. A dune barrier was constructed between the islands by installing appropriate sand traps: rows of reeds and twigs and plantations of marram grass.

In spite of the massive effort, the result was an uneasy compromise characteristic of all humanity's attempts to master the elements. The dunes, mobile by nature, continued to break through the constraints imposed by society. Every so often, further curtailing was needed.

Today this landscape is used for recreation, and as an important source of drinking water for the densely populated, sheltered lowlands behind. "With today's technology we have the dunes in the palm of our hand, for the first time in history," van der Meulen says. "Curiously, this gives us the freedom to handle these terrains much more liberally than before. In restricted areas we can even allow the dunes to move about freely."

Vlieland itself (figure 5.1) is also the scene of an age-old struggle between natural and cultural forces. In nature, such islands do not stay put: they shift with the waves and the winds. The string of islands to which Vlieland belongs has been moving for ages towards the mainland, despite all attempts to keep it in place. Long ago, there were two villages on this particular island, West Vlieland and Oost Vlieland. The former settlement has been swallowed by the sea, and now only Oost Vlieland remains, a lovely holiday resort. The remainder of the island consists of the beach, the dunes, and a few small patches of grassland. And although at strategic points breakwaters have been installed to protect the land from storms and high seas, here you can feel as free as the wind. No motorized vehicles are allowed on the island—cycling is the means of transportation along the various paths. It is a paradise for quiet families with small children, amateur naturalists creeping over the ground with botanical cases and cameras, and booted fishermen in the mud flats. Here, by its absence, do you become aware of the stench of our cities. Instead, here there are the flowers, the birds, the air.

A walking tour of Vlieland's mobile landscape vividly reveals the interplay of biological and nonbiotic forces. One can see the effects of living things on moving sands while studying the rock cycle in full swing as it operates at the surface all around the earth. An incredible variety of climatic regimes is represented on this island. On the west side, for instance, north and south beaches widen and meet to form a Dutch Sahara, the Vliehors. The flat desert land-

5.1 Map of Vlieland.

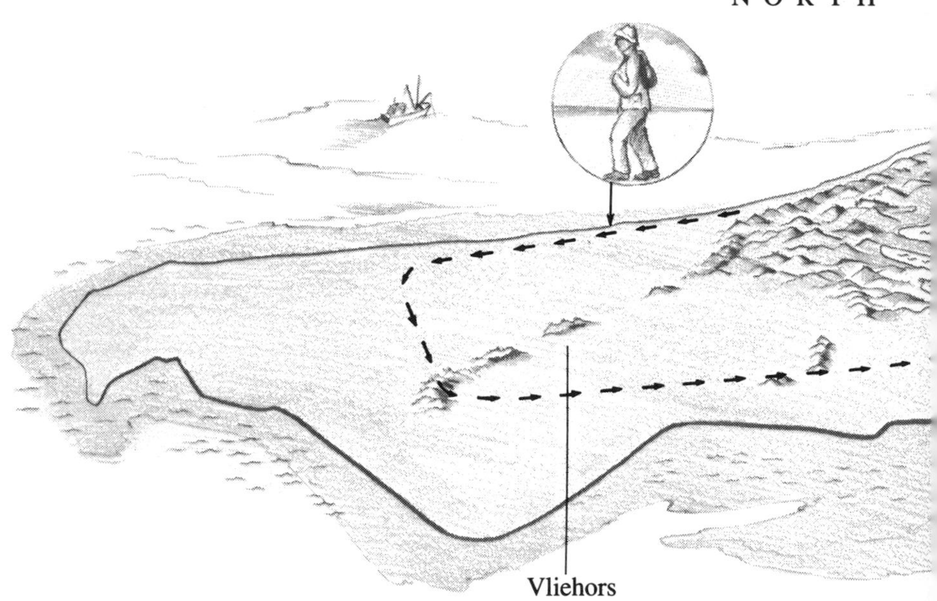

scape is broken only in the middle by a few isolated dunes, the abandoned remains of a man-made sand dike. If you walk inland about 500 yards, the sand is a white ocean blazing in the sunshine. The sea, inaudible now, seems a mirage.

Microbes, Building the Land

The earth's surface here is smooth. This is not the usual loose beach sand, which makes walking a challenge. Instead, the sand seems fixed by some enigmatic glue. The ground is slightly damp, although it is well above ground-

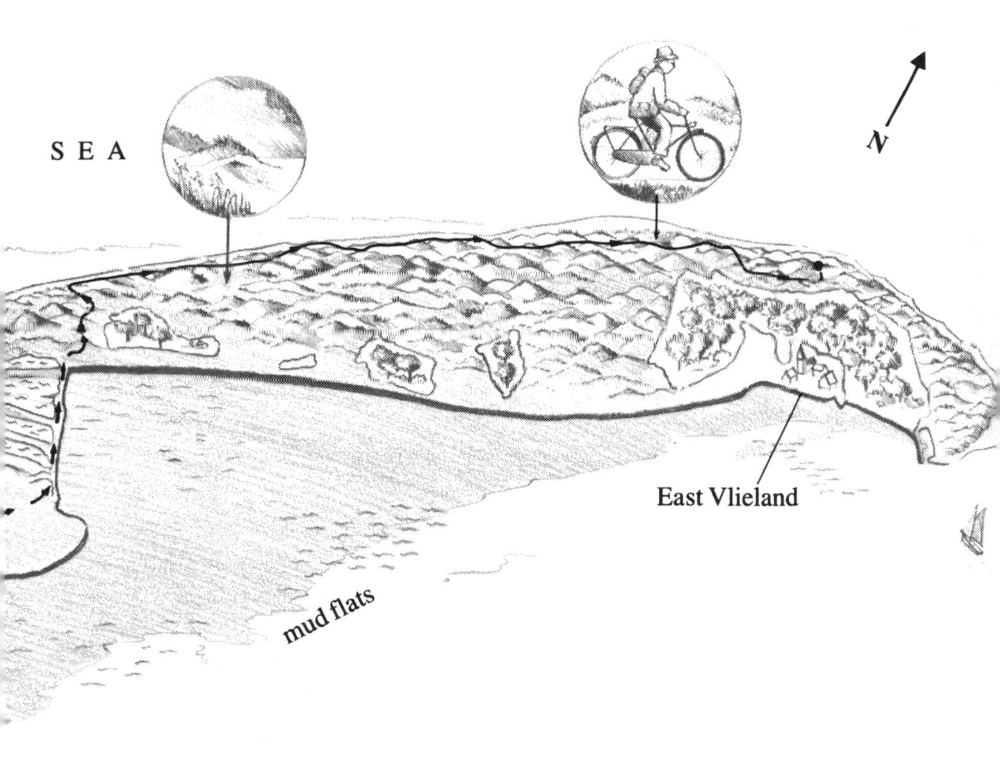

water level. Feet barely compress the sandy surface, but the vague imprints become curiously green, as if enriched by some magic fertilizer.

If you kneel and remove some of the surface with your finger, the first thing you notice is that you can peel off the top quarter-inch of the sandy sediment. The material is moist and has some internal cohesion. Somehow, despite the blazing sun, it is able to retain water. In contrast, the sand below is loose and dry. If you take a little slab of the hardened top layer and break it in two, you see a distinct lamination with colored bands parallel to the sedimentary surface. A very thin layer of sand on top, then green, then purple, and finally black. It is a stratified microbial community known as a *microbial mat.*

Microbial mats are composed of a fine, complex tissue of bacteria, living and working in the sand. They occur in hot springs, salty pools, deserts, and here, on nutrient-poor sands that are occasionally invaded by the sea. They thrive in harsh and inhospitable environments where few other organisms can survive. And, despite their modest appearance, they are among the most fascinating of all living systems.

Through a pocket lens one can see that the sand in the green layer is held together by very fine green filaments. The filaments are intertwined, and their sticky surface keeps the sand grains in place. These are cyanobacteria (or blue-green algae, as they used to be called). The filaments are long strings of bacterial cells surrounded by a sticky organic sheath. The green pigment is chlorophyll, very similar to the substance that stains the leaves of plants. It plays an essential role in photosynthesis, the process whereby sunlight is captured and then converted into chemical energy of a form that can be used to fuel all the cellular activities.

Cyanobacteria are an amazing group of creatures. Most of them are extremely self-sufficient. They can live in water either fresh or salty, survive prolonged periods of desiccation, and glide around in the sand until they find the most appropriate site in which to live. They make their entire microcosm from sunlight and simple minerals. With their photosynthetic apparatus they can produce oxygen; it is believed that they were the first organisms on earth to introduce this highly reactive gas into the atmosphere. They capture carbon dioxide and nitrogen from the air and convert these gases into organic matter. They can efficiently scavenge phosphate, another essential nutrient often in short supply, and store it for use during periods of hardship. They can dissolve mineral grains of many kinds, and extract the nutrient elements. It is safe to assume that the cyanobacteria form the backbone of the entire microbial community in the mat.

The purple layer below is dominated by photosynthetic bacteria of a different kind. With their red-pigmented photosystem, these bacteria capture some of the light that chlorophyll cannot absorb. They neither produce nor utilize oxygen. Instead they convert sulfide to sulfate. Finally, the black layer at the bottom represents yet another regime. It is inhabited by a brand of bacteria that makes sulfide out of sulfate. The sulfide is either used by the red microbes on top, or reacts with iron to form the iron sulfide mineral that gives this layer its deep black color. This is how Garrels's pyrite reservoir is constructed (chapter 4).

Together these organisms form a dense, highly structured community of interdependent species stretching out over the sandy plain as far as the eye can see. The community is self-supporting: it diligently collects energy from the sun, extracts nutrients from the environment, restrains moisture, captures sand—and thus creates and maintains the complex microenvironment it needs in order to thrive.

At the shoreline, the wash of the waves produces ripples on the sedimentary surface. These ripples penetrate far into the sandy plain, and loose sand accumulates, transported grain by grain. But where microbial mats grow, no ripples occur. Only blocks of ice or wood break the flat surface on the rare occasions that the sea reaches this area; the waves themselves remove little sand. Rather, the microbes capture loose sand, gliding up and pasting the surface together. The whole community then follows, and soon the land has been raised by a tenth of an inch or so. Microbial mats are pioneer communities that play a distinct role building and stabilizing islands such as Vlieland.

If too much sand is added at once, the mats cannot cope. The bacteria die or wait until the wind removes some of the sand. On top of the flat mat-made surface are numerous miniature sand dunes no more than 30 feet long and a foot high. They are sickle-shaped, with beautiful, regular outlines and crests as sharp as a knife, their points facing downwind. Such elegant accumulations are typically formed by the interplay of wind and fine sands. No organisms take part in their formation. And yet, would these dunes be the same if the microbial mats did not tie up the bulk of the sand, providing a firm and flat substrate?

What a curious landscape this is. Sand all around, virtually flat—and yet, one can distinguish three zones. On the shore and in the irregular channels that meander inland, the sand has been modeled into regular ripples by the waves and currents of the sea. Then come the low, crescent-shaped sand dunes. Both are purely physical features, produced by the dragging of different media over the sand, water in one case and air in the other. But between the two, and underlying the crescents, lies a biological construction: a thin, perfectly smooth, horizontal slice of bacteria, only a fraction of an inch thick, and many square miles in extent.

Approaching the Dunes

If you look up at the terminal patch of sand dunes, still half a mile away, what a striking contrast. Not only are these hillocks much bigger than the modest sickles at our feet, but they also lack their geometrical outlines. Plump and irregular, they rise abruptly from the plain, like collapsed plum puddings.

As one approaches this strange new landscape, the transition seems less abrupt. The sand sickles become higher and larger; they coalesce into elaborate festoons, and then come the first isolated tufts of grass, each standing out amidst a small heap of sand. These little pinnacles are distributed around the main hillocks, and the grasses on them are outposts of the army of sand binders that inhabit the dunes by the millions.

The ingenious strategy deployed by these grasses in trapping and pasting the sand was explained to me by Gerrit Jan de Bruyn, a biologist and a colleague of mine at the University of Leiden. The lank leaves of the grasses stand out in the wind, forming impervious, flexible screens that locally moderate and inflect the air flow, and encourage the suspended grains to fall at their base. Then fresh sprouts grow and catch more of this valuable dust. While the grasses work themselves upwards through the substrate that they have collected, a complex network of roots links these tufts of grass underground. Perfect rows of new roots and plants spread from the subsurface network, like tentacles carefully sensing the colony's surroundings, ready to take over the territory and extend it. All the while, the vertical growth of the grasses keeps pace with the accumulating sand. When the roots are buried too deeply they die, and a new network forms higher up.

At the foot of the main sand dunes, wherever the sand grains blow up to a level where the grass can settle, the grass enhances the capture of sand. No longer will the physical forces of the sand-loaded winds determine the shape of the accumulations. These are vegetative colonies of irregular outline, kept in place by the substrate they collect. The small mounds are exceedingly vulnerable, however. A single incursion of the sea, or a vehement storm, can easily sweep them away. The destruction inflicted by such events is obvious: deep gullies slash the grass-covered sandy barriers. In this area of transition between the sandy plain and the elevated dunes, brutal physical violence and the soft, steady workings of biology compete for mastery over the sands.

Living systems rarely, if ever, take possession of their environment by brute force. The impact of life forms is subtle; where overwhelmed, they give way. But as soon as conditions permit, the living systems return, modulating the circumstances, orchestrating the physical forces to their own advantage. Thus they construct the dunes, and this whole island.

On a Lonely Dune Top

When one perches atop the highest dune, one immediately senses that a critical threshold has been overcome. The grasses have managed to master the storms and the waves of the sea, and together their legions have raised their substrate 30 feet and more—a safe distance from the occasional incursions of the salty waters below. Compared with the humble pinnacles at their feet these dunes are enormous constructions, organized living systems made out of grasses and sand. But even these high barriers are transient features. Their sands are being churned up all the time, the grass unearthed and the elaborate framework of roots destroyed. Life and death go hand in hand in this land of motion.

Marram grass is the dominant species here, proudly waving its high plumes as if defying the wind and the sand. Not the prettiest of plants, but this tough pioneer exemplifies better than anything else the geological force of living things (fig. 5.2).

Marram grass is superbly adapted to this inhospitable environment. At regular intervals it is exposed to intense sand-blasting. On warm summer days, the sun dries and heats the sand. This place can be a desert, albeit a small one. No small organisms can live here without special adaptations. The dryness forces the stiff, spiky leaves of the marram to roll up, and the stomata to close off tightly. But when the air is moist, or in morning dew, the blades unroll and their insides, with regular stripes, are exposed. Then the plant can transpire and exchange gases with the atmosphere. Marram is bluish-gray. This is the characteristic color of all desert vegetation, for blue-gray reflects the heat of the sun, and helps to cool the environment.

In only one respect does marram grass appear to fail. It grows in small clumps, leaving the yellow sand in between uncovered. It looks as if it is challenging the elements, inviting them to blow out the loose substrate. Yet, there is a reason for this. Marram grass is at its best in the dunes at the shore and just behind, where the sand is churned up continuously. It thrives by scavenging the scanty nutrients in the blown-up sand, and would die if the storms and the sea stopped bringing fresh supplies to the surface. The loose sand and the churning motion are just what this plant needs. Farther inland, the grass looks rather miserable. Its

job is done: it has built its barrier dunes, paving the way for other communities farther from the shore and less dependent on the wind for nutrient supplies.

Living things do not simply adapt to the vicissitudes imposed by the physical and chemical environment. They

5.2 Marram grass in the dunes.

also modulate the environment, actively adapting it to their own needs. They even assist the elements in their destruction when this enhances the community's chances for survival.

On the Lee Side

On the lee side of the island on the southern shore, the sandy plain merges into the mud flats of the Wadden Zee. This is a different world, with its own physiognomy, sedimentary structures, vegetation, and animal life.

The first sign of change is that the microbial mats become stronger and denser. Locally, immediately behind the dunes, they may form a coherent, green, and tough organic layer. These are mats in the familiar sense. You can peel big slabs off the substrate. When you tear one apart you can see the fine tissue of cyanobacterial threads.

Plants, including the tasty, salt-loving samphire, grow on these sands. And algae flourish, staining the surface with their bright greens. At low tide birds by the thousands feed on the invertebrate animals that crawl around in and on the exposed mud shoals. The sand beach closes in here, leading to plant-covered mud fields. This vegetation is voluptuous and varied. Its distribution follows a distinct pattern that parallels the coastline and is plainly related to the ability of the plants to tolerate occasional incursions by the sea. The samphire grows nearest to the sea, whilst dense reeds occur next to the dunes where fresh water seeps in from the high sands. In between these extremes live plants adapted to intermediate levels of salt.

The mud fields are interrupted by creeks worn out by the streams of the tides. A brown, slimy substance is abundant in these creeks—a muddy sediment, rich in clay, and to which nutrients can easily stick. The slimy cover is made by diatoms, microscopic unicellular algae encased in delicate skeletons of silica. The shoals, which are covered with water at high tide, are entirely swathed by the diatom mats.

Some years ago, Poppe de Boer and Peter Vos, geologists at the University of Utrecht, conducted an experiment in an estuary in the south of Holland. Huge masses of water move in and out with the tides there, and the sediment is thrown up as high as 30 feet. It had always been thought that these sediment masses move freely in the strong currents of water, but when the scientists killed the diatom mats on one of these banks with a copper sulfate solution the effect was dramatic. Within a few days, the bank was eroded away while its neighbors held fast. There was no doubt: diatom mats, like dune grasses and microbial mats, keep their substrate in place by trapping suspended particulate matter.

I must not forget yet another brigade of these land-making legions. Billions of cockles and other invertebrates crawl around in this mud. They continuously burrow through the sediment, and at high tide they filter the water on top, capture particles that had been freely suspended, wrap them up in their slimes, and then spit them out, in big lumps, ready to be glued to the growing sediment mass.

Riches of the Soils

Further inland, on parts of the island more protected from ravaging storms, plant cover is richer, flourishing on a layer of soil no more than 10 inches thick. In comparison to the sand that so sparsely supplies nutrients to the marram grass, the soil has been fixed for a long time, maybe for hundreds of years. How could the much richer vegetation of these soils have survived for so long without access to fresh sources of underground nutrients? The answer is simple. The soil itself is the nutrient source. Soils are not just mixtures of roots and sediments; they also contain elaborately structured communities of bacteria, fungi, insects, worms, and other organisms. These diligently clear and digest the dead bodies of plants and animals, and release their nutrient contents, which then can be reutilized by the entire community. So, the soils and vegetation are complementary. The sunlit parts of the plants capture the energy that fuels the entire system, and the microbes in the soil recycle the nutrients. Together, they form a stable community that can survive for hundreds of years even if supplies of nutrients from outside are minimal.

Intricate patterns of living communities spread out over this fleeting land of beauty, plastering the sand on the surface and keeping the island in place. It is clear that their distribution is by no means haphazard. Successive communities follow one another until finally the most stable system prevails. This succession depends on microbes which bring nutrients (particularly minute quantities of nitrogen and organic carbon) into the ground, ever enriching the soils so that they can support a more intense growth of plants: lichens, mosses, grasses, and finally, flowering plants.

Yet, the dunes do not last forever. The cycling is always a bit leaky, and nutrients are lost all the time. Eventually, the soils are depleted. The impoverished vegetation is an easy target for the autumn storms. The dunes shift, the sands turn over, and the story begins again. Left on its own, the whole island would drift with the waves and the wind. Life forms slow down the process, but they certainly don't stop it. The natural plowing of the ground is essential: it brings a continuing stream of foodstuff to the surface and keeps the world alive.

The Power of Living Things

The question is, what do life forms have to do with the cycling of rocks? The obvious targets of biological intervention are the processes going on at the outer earth: the cycling of water; weathering; soil formation, erosion, transport, sedimentation, and finally, lithification. This day in Vlieland we experienced the immensely rich repertoire of the power of life, but is this representative for the earth as a whole?

The Water Cycle and Climate

Water is the medium that carries rock debris from the mountains down to the ocean floor. It is evaporated from the seas, blown over the surface of the entire earth, precipitated, and carried downslope back to the sea. Without water no life can exist, and there can be no weathering, no washing of the rocks to the sea. Is water influenced in its circuitous course by the biota? We have seen how microbial mats can retain some moisture in their slimy tissues, even in the full blaze of the sun; so do soils and plants. Some of the water that percolates down through the permeable dune sands is held by the roots and the microbes—stored until the rain delivers new supplies.

Plants transpire. They pump large amounts of water from the ground and allow it to evaporate from their leaves. This upward stream of fluid helps them remain cool and provides them with nutrients from below. Transpiration takes place at a massive scale, most of all in tropical rain forests. It sustains cloud formation and rainfall, and activates the cycling of water on land. Computer models suggest that through transpiration plants exert a massive influence on the global climate. In the absence of vegetation, extreme dryness would prevail over large expanses of the continental surfaces that now harbor a lush vegetation, and summer temperatures would be 15°C to 20°C higher.

Life does not stop or induce the water cycle. It appears, by transpiration and other mechanisms, to modulate its operation, to divert, intensify, and exploit it. This massive intervention drastically changes the earth's environment, and favors the conditions that maintain life on this planet.

Weathering and Soils

In Vlieland you have only to go to the graveyard and inspect the stones to see the devastating evidence of microbial handiwork. This is weathering—the breakdown and decay of rocks. Polished rock surfaces quickly lose their luster, and where graves have been neglected for a couple of years you can see the mineral grains disintegrating. Fungi, bacteria, and lichens of all kinds thrive on the gravestones; they break them apart, scavenge the nutrients, and spread out over the surface, displaying their rich colors. Later, mosses take over and, eventually, vascular plants: grasses, herbs, and shrubs. They all feed on the nutritious minerals within the stones, leaving barren sand grains and clays. We should spray our monuments with antibiotics to prevent them from being eaten.

Microbes and plants are miners. They collect raw materials from the solid surface of the earth and make them available to all other organisms. This miner community captures nitrogen and carbon from the air, converts them into organic materials, and stores them in the rocks. Some of these organic compounds attack the minerals and capture valuable metals.

Soils form, little by little: vulnerable, friable surface layers, highly structured communities of microbes, fungi, roots, worms, and beetles, closely associated with a wealth of different minerals, moisture, and macromolecular slime. It is in this microcosmos, evolved over billions of years, that the nutrients are recycled and circulated.

Erosion and Transport

Living systems adapt rocks to their own needs. The sticky soil layer once was rock, and was made fit for life by life itself. It is touching to see the vast repertoire of mechanisms that has evolved over the eons to keep this nutritious, living substrate in place. The hard rocks are destroyed, but the fragile soils are preserved, by glues, roots, water, and organics. Is not the whole island of Vlieland a monument to the perennial struggle between the conserving forces of life and the destructive forces of wind and water?

Figure 5.3 demonstrates the mitigating impact of life forms on erosion and transport. The data were collected in the western United States. The figure shows the amount of transported debris plotted against the mean effective precipitation. In the Utah desert, where both rainfall and vegetation are low, transport reaches a maximum and decreases with increasing precipitation until it reaches a stable level in a forested area. In the absence of living things, the curve would rise higher and higher, and no maximum would ever be reached. Think of the dust bowl that scourged the United States in the 1930s, and the sahel today, and one realizes what forces are released when the biological protection of soils is destroyed.

Yet, there is another side. As I mentioned before, the

cycles run by living things are leaky, and valuable nutrients are lost. They are washed out bit by bit, or dissipate into the air. Soils mature and then become senile, unable to support further life. By instinct we may feel that this is a shortcoming of the biological organization. But it all works out rather well. The lost nutrients become fertilizers downslope or downwind. They spread out over the continents, and eventually enter the sea where they support life in the waters at

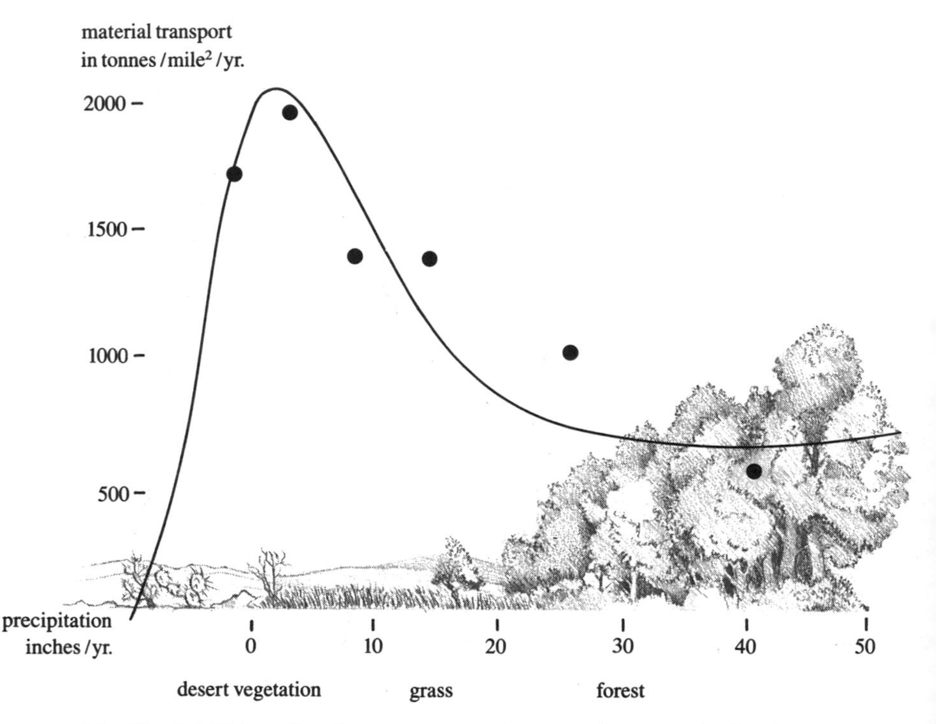

5.3 The inhibition of sediment transport by vegetative cover. In the absence of vegetation, the amount of transported material (*vertical*) would continue to increase with increasing precipitation (*horizontal*). The vegetation cover becomes more dense with increasing precipitation, however. This holds the rate of transport to a low and stable level. *Based on S. A. Schumm,* The Fluvial System *(New York: Wiley-Interscience, 1977). Reprinted by permission of John Wiley & Sons, Inc.*

large. Like marram, soils are transient phenomena. They come and go, and when they are gone, a fresh supply of rocks always remains to be eaten (at least as long as plate tectonics continues to do its job). Life cannot survive without destruction.

Life and the Sediments

Wherever you look, life forms scavenge particles, dissolved materials, and gases from the environment, and then sort and add them to the sedimentary column. The biological intervention in sedimentary processes occurs at a massive scale. Gigantic amounts of carbon, for instance, are stored in the rocks. Compared with these huge reservoirs, the amounts of carbon dioxide in the atmosphere, bicarbonate in the oceans, and organic carbon in the living biomass are minute. The concerted activity of the biota is responsible for or at least helps the production of these masses of carbon in the rocks, of which limestone and other carbonates are the major constituent. Coral reefs, carbonate platforms such as the Bahamas, lime oozes in the deep sea—it is now clear that organisms have accumulated virtually all of this material. After rocks, the largest carbon reservoir is made of organics: remains of the cells and organisms converted into gas, coal, oil, and less well-defined materials distributed in the sedimentary rocks.

Living things make rocks that would not form in their absence. Some diatoms inhabit the mud flats, as we have seen, but others float around in the ocean, blooming in places where nutrient-rich deep waters well up to the surface. They scavenge silica from the water and reprecipitate it to form intricate opal skeletons. As a result of this activity, seawater itself is almost entirely depleted of silica; below the upwelling waters, however, thick deposits of diatom shells cover vast areas of the deep ocean floor. Many valuable mineral deposits are formed through biological intervention. Manganese, iron, copper, uranium, and gold are a few of the metals that can be scavenged by organisms from surface and ground waters and accumulated in the sedimentary column.

Life and the Rock Cycle

Figure 5.4 (A) shows the rock cycle in its simplest form. Rocks are carried from the elevated continents towards the deep sea, and are then returned by plate tectonics to the mountain ranges. Is it possible that the extinction of all life on earth would bring the operation of this basic mechanism to a halt? Although the evidence given in this chapter in no way supports such a contention, I would not be surprised if future research revealed that living things were necessary for the maintenance of plate tectonics.

For the moment let us assume that the rock cycle does not depend on living things for its basic operation. What, then, is the impact of life? It is fair to conclude that the biota subtly but profoundly modify the outer part of the cycle—the flow of rocks from the continents down to the oceans. Weathering is speeded up and intensified, sedimentation is catalyzed in innumerable ways, and erosion and transport are strongly inhibited. Debris is diverted on its way down the slopes, and its various constituents are conducted along pathways that would not exist in the absence of living things. But the basic mechanism does not seem to be affected. This might be one reason that traditional geological thought has neglected the impact of life for so long. The other, perhaps more important reason, is that despite the obvious geological effects of life forms in the recent environment, we lack the theory by which most of these effects can be recognized in the rocks.

Figure 5.4 (B) gives an alternative, biological view of the cycle. It provides a simple overview of the function of its operation for the maintenance of life. On the continents, nutrients are extracted from the rocks. But the supplies brought up from the earth's interior are so small that they have to be recycled to sustain life's frenzy. The biological cycles leak, and so, very slowly, the nutrients move down to the oceans. Eventually, the nutrients are dumped on the floor of the deep ocean, from whence plate motions carry them back to the subduction zones and they descend to the furnace of the deep earth. There they are intensively processed, and the original foodstuff is reassembled for use by living things once again.

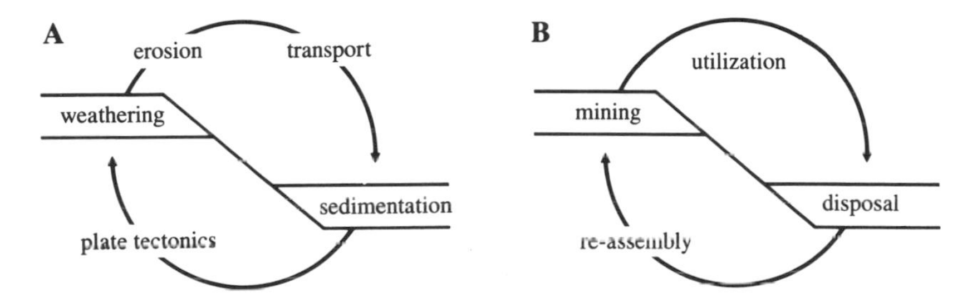

5.4 (A) The rock cycle from a physical point of view. (B) The rock cycle from a biological point of view.

6

THE POWER OF THE SMALL

EVERY MINERAL has a characteristic chemical composition, its constituent molecules neatly arranged into a three-dimensional lattice—a crystal. As a rule the orderly structure of crystals is reflected in their regular, multifaceted geometry; sometimes, however, the minerals boast highly fanciful forms. This may reveal the indelible fingerprint of life. Piles of these minerals—refuse of the dead bodies—become massive geological features, rocks and cliffs.

That very small organisms can shape crystals is easily shown. Take two glass beakers and fill them with water. Add calcium chloride to one beaker and sodium carbonate to the other, and stir. The powders will dissolve like sugar in a cup of tea. Now slowly pour some of the sodium carbonate solution into the calcium chloride solution. After a while a white cloud of calcium carbonate will form, or precipitate, from the dissolved substances. Figure 6.1 *top* shows this precipitate at high magnification. It consists of tiny crystals of approximately the same size with almost perfectly lozenge-shaped surfaces. This type of calcium carbonate is calcite, the major constituent of limestone and one of the most abundant minerals on earth.

Figure 6.1 *bottom* also shows calcite but here the mineral is organized and curtailed, subjected to the stringent demands of a living cell. A large number of wonderfully shaped, interlocking scales called coccoliths form a loose casing surrounding the globular cell of a single-celled alga. The name of this organism is *Emiliania.* The cell is a hundredth of a millimeter in diameter—if you lined up 5000 of them, they would measure about an inch. *Emiliania* is a kind of plankton, a single-celled marine organism that floats freely in the upper layers of the open ocean where it collects the solar radiation necessary for its energy.

The photographs depict the distinction between a mineral and a biomineral, the worlds of physicochemical and of biological forces. The question is, does the calcite of the earth's crust belong to the world on the left or to the one on the right?

How Emiliania *Makes Calcite*

To answer this question, we have examined how *Emiliania* molds the calcite crystals. Somehow, the alga exerts a high degree of control over the crystallization process. Indeed, the calcium carbonate of *Emiliania* is carefully nurtured inside the cell in a minute specialized cavity, or vesicle.

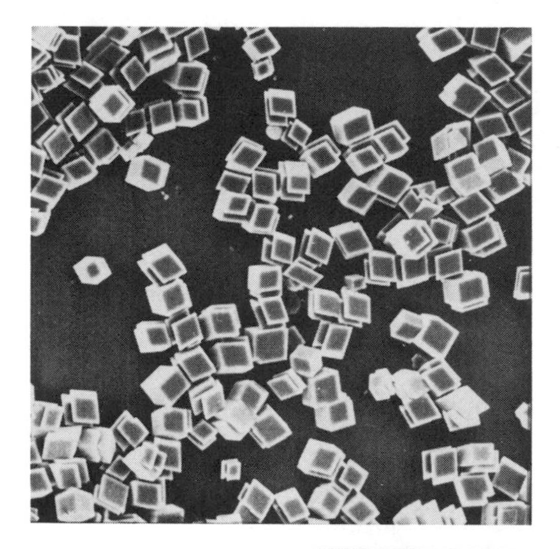

6.1 (***Top***) **Crystals of the mineral calcite, $CaCO_3$.** *Courtesy Dr. P. R. van Emburg, University Leiden.* (***Bottom***) **One cell of *Emiliania huxleyi* covered with coccoliths (oval scales of the same mineral, calcite). Each coccolith consists of a radial array of elaborate crystals. The coccoliths are formed inside the cell and extruded after their crystallization is terminated.** *Courtesy Dr. Annelies Kleijne, Free University, Amsterdam.*

There its growth is programmed with great precision, so that the same extravagant structure is produced over and over again, billions of times.

In my biochemistry laboratory we have spent many years finding out how this is done. The important thing to remember is that coccoliths are more than plain calcite. Attached to the mineral is a polysaccharide, a giant molecule built up from a large number of sugar-like components. This is one of the innumerable biochemical instruments of great sophistication and delicacy that organisms consist of and upon which their organization depends. This particular polysaccharide, formed within the algal cell, is one of the most complicated ever described. Despite the fact that even under an electron microscope it can hardly be seen, its workings are as complex as a television set or an automobile.

Why are these giant molecules attached in vast numbers to the coccoliths? We discovered that the polysaccharide has a distinct effect on the formation of calcite. When we repeat the precipitation experiment, this time adding a minute amount of the polysaccharide to the solution, we find that no calcite crystals will form. Rather surprisingly, the polysaccharide does not aid crystallization—it inhibits it. Probably, it does this by binding to the crystal nuclei and enveloping them, preventing the nuclei from growing. The cell can use this property to regulate coccolith growth. Each one of the radially arranged coccolith elements can be viewed as a single crystal, the growth of which is allowed to proceed only in certain directions. Somehow, the polysaccharide is kept away from the growth fronts and plastered against the surfaces where growth is prevented.

Every two hours as long as the sun is shining, a new coccolith vesicle is produced. A hollow space is created in its interior from which the polysaccharide is pulled away (figs. 6.2, 6.3). Immediately, very small calcite crystals are formed along the rim of a preformed oval organic plate. By some unknown mechanism they are perfectly oriented from the start. Now all the cell has to do is allow them to grow out in selected directions, while repressing the growth in other directions. Crystallization is tamed, and subjected to the rigorous demands of the cellular organization. A biomineral is formed. It is good to remember that even after many years of intensive research we don't know precisely what happens. Unfortunately, it is not the job of university officials to be inspired by beauty and mystery, but to take that the show can go on. They want efficiency. They see a geologist in a biochemistry department as an irregularity, an ideal target for cutbacks. And so, our work is watched with suspicion, and time upon time our research group has had to battle for survival. Somehow, we had to show that *Emiliania* is not only interesting but important.

It is an easy point to argue with conviction, because this tiny organism plays a role in earth dynamics far out of proportion to its modest size. Figure 6.4 is another representation of *Emiliania,* seen not through an electron microscope but from a satellite in space. One can easily recognize the northern part of Scotland and the islands of the Outer Hebrides. The vast cloud in the ocean is *Emiliania.* Under appropriate conditions, these cells can grow explosively, forming giant blooms thousands of square miles in size. Such transitory but recurring and gigantic accumulations occur in all the world oceans. *Emiliania* is believed to be the most productive calcite-producing species on earth. It also is an important element of the oceanic biota, with a prominent role in the food webs.

The cells and coccoliths are so small that, given the turbulent motion in the seawater near the surface, they might remain suspended forever if left on their own, even long after their death. But when a bloom of *Emiliania* dies off, the cells clump together so that they sink rapidly through the water. Or they are eaten by small animals, especially by a tiny swimming crustacean, the copepod. These creatures digest the soft parts of the algae, wrap the coccoliths in a slimy substance, and excrete them as fecal pellets. The pellets are biological bullets, loaded with coccoliths and other small particles, that shoot down, reaching the ocean floor in only a few weeks. That is how coccoliths accumulate; they are produced and transported by biological forces.

Enormous carpets of coccoliths, larger in area than all the continents together, cover the floors of the world oceans. These terrains represent the largest calcium carbonate sinks on earth. The tiny coccolith-generating alga that we can observe with the electron microscope is geological dynamite, a formidable force helping to conduct vast fluxes of calcium and carbonate toward the ocean bottom.

Numerous organisms produce calcium carbonate. For instance, the corals, mollusks, algae, bryozoans, and sea urchins that make up coral reefs and inhabit our coasts all have hard shells made of this material. Their mineral skeletons are a heavy load, forcing these organisms to live on the floor of the ocean. They are concentrated on the continental shelves where the supplies of light and nutrients are optimal. What is so wonderful about *Emiliania* is the open, delicate shape of its coccoliths. The lightweight skeleton does not prevent *Emiliania* from assuming a floating lifestyle, and moving out into the upper domains of the entire ocean. If its mineral coat were to become bulkier it might

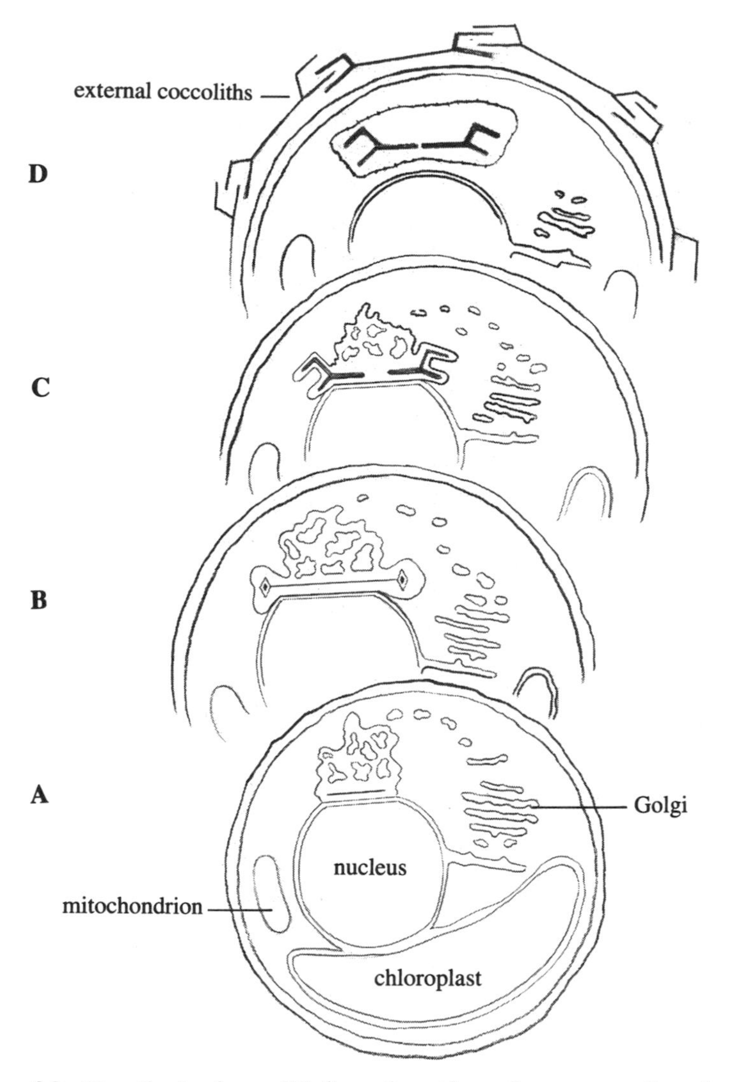

6.2 Hypothesis of coccolith formation. Above, four cross sections through a cell with—from (A) to (D)—successive stages of coccolith production. The coccolith forms inside a special vesicle that adheres to the cell nucleus. The detailed cross sections in the illustration on the facing page show the thread-like polysaccharide molecules lining the interior of the coccolith vesicle, like the hairs of a fur. The fluid inside the vesicle is supersaturated for calcite, while the polysaccharide inhibits crystallization. (A) Initially, a thin organic scale is formed. (B) The vesicle begins to dilate at its periphery, pulling the polysaccharide away from the margin of the organic scale. Immediately, a neat row of

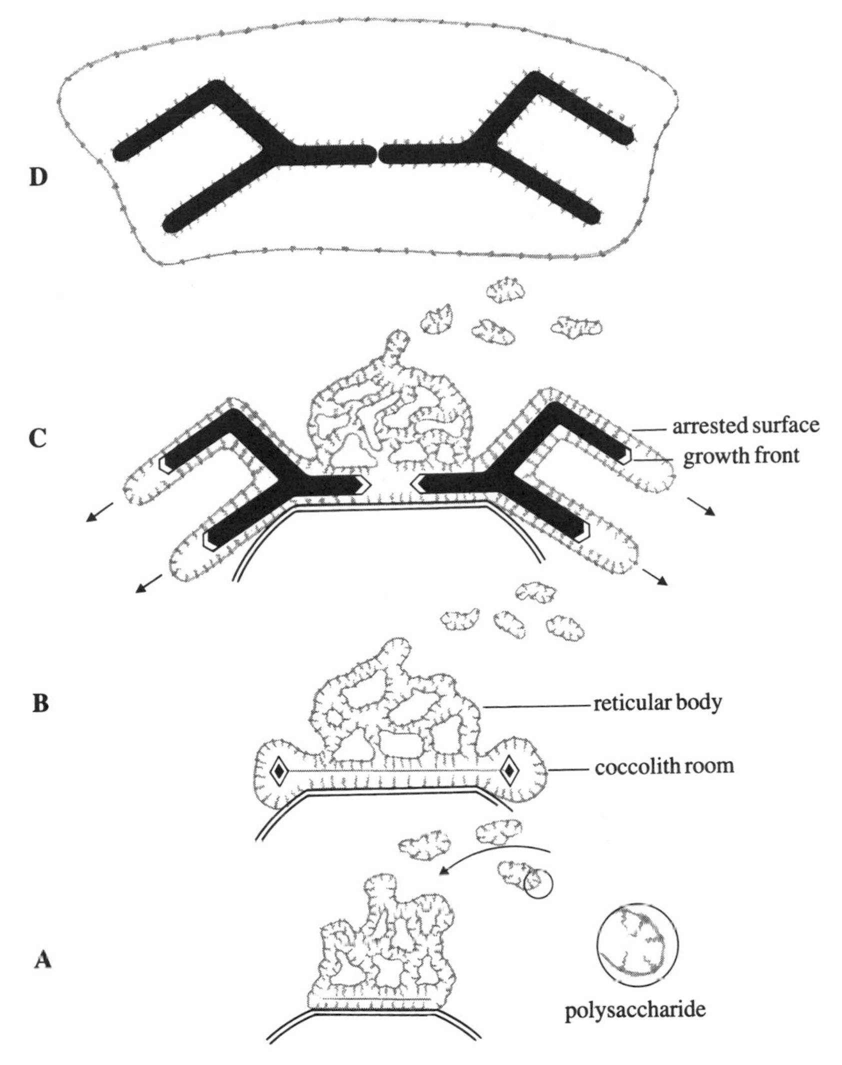

crystals is formed inside the hollow space. (A three-dimensional representation is shown in fig. 6.3.) (C) The coccolith vesicle now undergoes a precisely programmed change of form, pulling away the polysaccharide from each of the crystals in three directions. Three crystal growth fronts are formed (*white angles*), whereas the polysaccharide inhibits crystal growth along the rest of the crystal surface. (D) Finally, the dilation of the coccolith vesicle is terminated, crystal growth comes to a standstill, the polysaccharide is removed from the vesicle membrane, and the coccolith is ready to be extruded to the outer surface of the cell.

sink towards the deeper recesses of the water and become deprived of its energy source, the light of the sun. The awesome precision with which *Emiliania* can control the growth of its crystals is essential for its survival.

Coccolith formation came about late in evolution. The earliest coccolith-producing cells lived slightly more than 200 million years ago. They were the first organisms to bring about a steady rain of calcium carbonate from the upper, illuminated zone of the ocean towards the deep sea. The evolution of a minute vesicle extended the world's calcium carbonate sinks from the shallow periphery to the main floors of the oceans.

I like to speculate about yet another plausible effect on the earth's dynamism. You may remember that new ocean

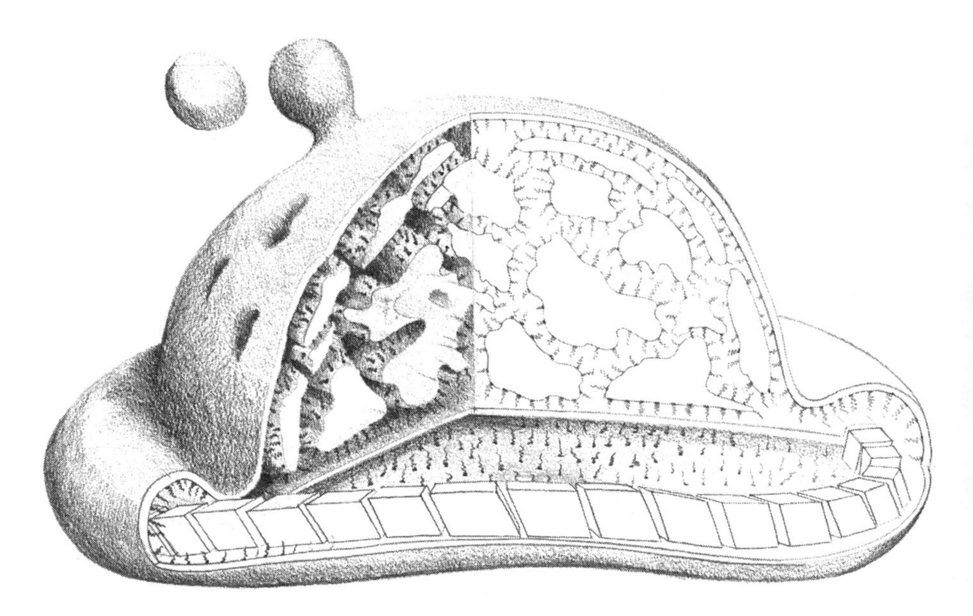

6.3 A three-dimensional representation of the process delineated in figure 6.2 (B).

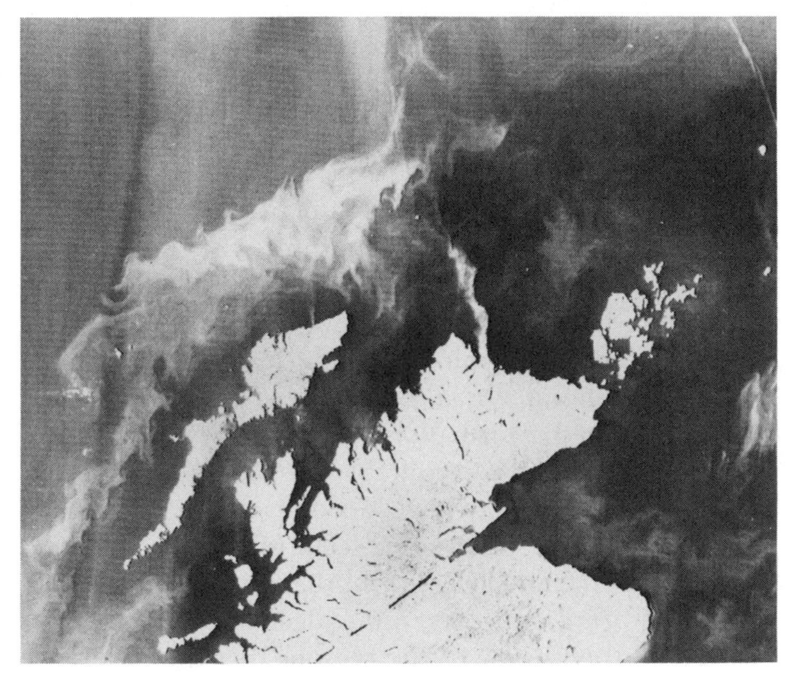

6.4 A giant bloom of *Emiliania huxleyi* in the Atlantic just north of Scotland, seen from space. No organism produces more limestone than this tiny cell. *Courtesy Dr. P. E. Baylis, University of Dundee, U.K.*

crust is formed at the mid-oceanic ridges, sliding away from there until, often hundreds of millions of years later, it reaches a deep-sea trough and vanishes into the earth's mantle. At the trough, most of the limestone sediment is scraped off the descending ocean crust but some part may be carried on into the earth's interior. There it is heated up, and enriches volcanic lavas with calcium while its carbon dioxide component is blown into the air. It is well known that carbon dioxide is a greenhouse gas, capable of retaining solar energy. The higher its atmospheric concentration, the warmer the climate. If evolution had not triggered the massive production of coccoliths 200 million years ago, much less of this gas would be around in the atmosphere, and the world's climate would be cooler.

Recently it has been pointed out that *Emiliania* is a potent producer of a smelly sulfur-containing gas, dimethyl sulfide. Blooms such as those in figure 6.4 are thought to be powerful sources of this gas. The fumes rise from the sea high up into the atmosphere and are oxidized by solar radiation into sulfuric acid. These minute acid droplets drift around in the air, forming ideal nucleation sites for the condensation of water. *Emiliania* is one of several algae held responsible for the formation of clouds over the oceans. Such clouds reflect a considerable proportion of the solar radiation, and they may have a cooling effect on climate. They are also thought by some scientists to induce a mild form of acid rain, and local storms that might stir the ocean waters and bring nutrients to the algal cells near the surface. Could this be one of the reasons that the fumes are produced? Both *Emiliania*'s coccoliths and gaseous emissions may have important climatic effects.

For decades, the biochemists used a single organism, *Escherichia coli,* to study the intricacies of cellular organization at the molecular level. It served as a model for all organisms on earth. One wonders if *Emiliania* will become the *E. coli* of the oceans, a model to study geosphere-biosphere interaction.

7

A ROCK EVOLVES

THERE IS HARDLY a place on earth where the eye does not rest on the geological workings of life. Yet the imprint of life on the rocks is not obvious. When we try to perceive the world as it was we see little more than the work of inanimate forces. *Emiliania* with its coccoliths is an exception. Indeed, in some rocks, limestone among them, the impact of life cannot be overlooked. The sciences of earth and life come together here. Let us now consider limestone in a broader perspective; our knowledge of *Emiliania* may help us keep the biological aspect in mind. We have within our grasp a new concept to aid our understanding of the earth's dynamics: the idea that rocks can evolve.

Sedimentology Meets Biochemistry

A few years ago I attended a meeting on limestones in Liverpool, England. For several days we had been bombarded with lectures on the structure and genesis of these rocks and of their ubiquitous remains in the geological record—the grey limestone masses underground, or uplifted and exposed in the landscape. The talks covered a variety of phenomena, ranging from the development of oceanic platforms and coral reefs to the minute individual crystals that make up chalk muds. I was well acquainted with this infinitely complex world from the time I had spent in the field as a geology student, but understanding of limestones had progressed greatly over the years. Intensive research, conducted by many teams and individuals, had revealed how the various sedimentary environments and the resulting rock types were distributed in the oceans and on land. The field had evolved from a static, descriptive science into one of dynamic interpretation. To achieve this result, modern geologists had mustered a whole repertoire of technologies including electron microscopy, sophisticated chemical analysis, and remote sensing.

Yet I felt that something essential was missing. The world of limestone sedimentation teems with bacteria and other organisms, slimes, and macromolecules. Every other lecture gave new indirect evidence of the overwhelming influence of biochemical systems on limestone formation. Speakers showed pictures of bacterial colonies and slime webs covering the surfaces and sometimes permeating the interior of the crystals. But nobody knew exactly what these microorganisms were doing there, nor how the study of their influence should be approached. A biochemical approach to limestone formation seemed to be badly needed, and our group had been working on that exact topic for years. As is the custom at such gatherings I had put up a poster (a written exhibit of our results), but so far had attracted little attention.

Then, in a pause between lectures, the man sitting in front of me turned around and asked, "Are you the man with the poster on coccoliths? I think it's fascinating, and I need to talk to you about it." He was Philip Sandberg, a geologist from the University of Illinois who is truly in love with carbonates, lime muds, and reefs.

That evening, Philip and I quickly agreed to collaborate. Coccoliths and limestone are made of the same mineral, calcium carbonate, of which calcite is one modification. Philip studies platforms of carbonate that often harbor huge masses of a kind of sediment known as mud chalks. It was his hope that with my biochemical orientation, I would help him look at these muds in an entirely new way. At the very least, we hoped to identify the biological entities responsible for producing the mud particles. He also yearned to know more about the slime that is so abundant where these sedimentary structures form. What is its biological source? Does it really inhibit the chemical precipitation of carbonate at the rate he suspects?

There was no reason for me to hesitate. With members of my lab I had pioneered the use of immunological techniques in paleontology, moving backwards in the geological record to study fossils of greater and greater age. When organisms die, their remains are scavenged by others and recycled, again and again. Yet a minute fraction has always escaped this fate, buried in the sediments for hundreds, even thousands of millions of years. Coal and oil are examples of highly concentrated accumulations of this material, but most of it is finely distributed throughout the rocks.

Together, this accumulated fossil organic material weighs more than 10,000 times as much as the living biomass on earth. Have these immense collections of macromolecules changed beyond recognition in the course of geological time? In an unconventional approach to this question, we have learned that with simple techniques we can detect, with very high specificity and sensitivity, the presence of macromolecules. Antibodies (specially tailored macromolecules) enable us to detect preserved remains among the smear of breakdown products. In some cases it is possible to compare macromolecules from different species of organisms and decide whether they were close or far apart on the tree of evolution.

Philip's deep involvement with sedimentology afforded a perfect opportunity for me to apply my work to geological research. After gathering chalk mud from a carbonate lagoon we could compare and study the distribution of the associated macromolecules in recent and fossil sediments. As with *Emiliania*'s polysaccharide, we could investigate their influence on the crystallization of calcium carbonate. Together, we agreed to explore the effect of biology on sedimentary processes using means other than the usual microscopy, and to see what was happening at the molecular level.

Grass in the Sea

"Here, Peter, look at this!" Philip stood waist deep in water and waved a handful of sea grass. This was not the introverted scientist that I had met in Liverpool. Now he looked rather like a pirate, in his diving outfit and with a huge red-and-white bandanna tied round his head. In the civilized world his full beard enhanced his appearance of respectability; here it emphasized his unreserved identification with the wild nature around us. This was Florida Bay: a vast, shallow lagoon behind the Florida Keys, bathed in sunlight, a limestone factory in full operation (fig. 7.1). Philip was completely covered with chalk, a fine, slimy, white mud.

I waded towards Philip and looked at the sea grass. Strange to find a flowering plant in abundance in the sea. Our movements had stirred up the carbonate mud from the bottom and now we stood in the middle of white, milky water. Carbonate is shorthand for calcium carbonate—limestone or chalk. The mud consisted of minute needle-shaped crystals of this material. Unlike clay, which is debris from crystalline rocks in the far-off mountains, the chalky

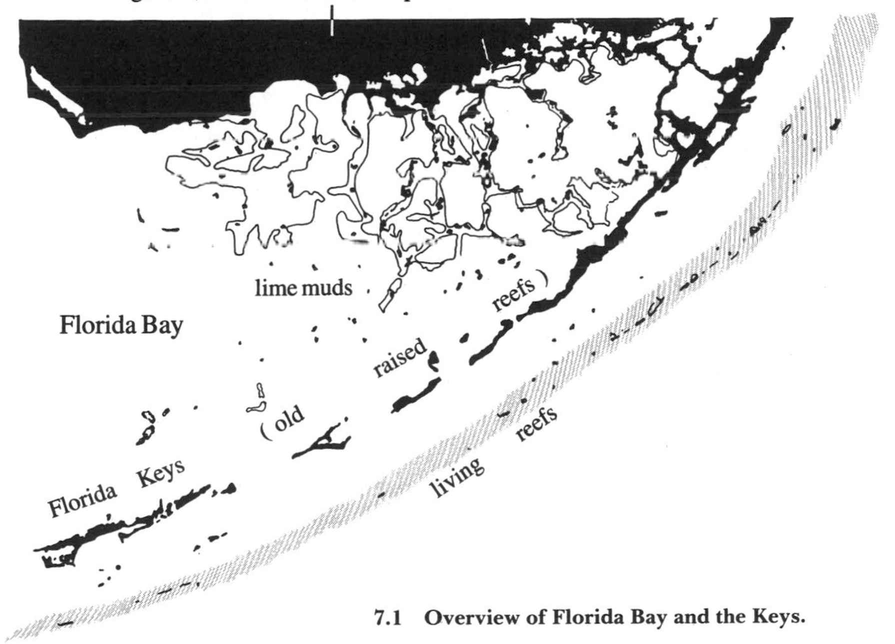

7.1 Overview of Florida Bay and the Keys.

particles were formed on the spot, in the seawater around us. We moved over to clear water and I could see that the grass leaves were long ribbons, about a third of an inch wide. Except for the youngest leaves, they were covered with a thin, white, irregular crust of chalk. On older leaves this crust practically obscured the green.

Philip explained that the grass leaves were covered with *Melobesia,* a small alga. "It may not look as if there is much of it, but the grass leaves are growing very rapidly, maybe six inches a month," he said. "New leaves continually emerge from the root and, after two months or so, the old leaves are shed." When these leaves die and disintegrate, he went on, the algal crust falls apart and a host of minute chalky crystals, contained in the tissues of the living algae, is released. That was the source of much of the chalk mud around us.

In essence, the grass ribbons Philip held in his hand were part of a living conveyor belt that accelerates the production of chalk mud. The vast numbers of grass leaves provide a huge surface for encrusting chalk-mud producers. They are continuously pushed up from the roots and then shed, so that they are actively replaced all the time. Of course, both components of the system—the grass and the calcified algae on them—depend on sunlight for their growth. They thrive in these shallow waters, protected from the turmoil of the open ocean, and they form enormous amounts of mud in a very short time.

I inhaled and quietly snorkeled away, gliding through the water like a lazy whale, looking into the wonderful, alien world below. Grass all over, fewer than 3 feet below me, waving in the currents, and covered with *Melobesia*. The leaves, up to 15 inches long, stood upright in the mud.

This vegetation plainly exerts powerful influence over

and above limestone production on its environment. Like the marram grass in the sand dunes in Vlieland, or like the mangrove brushes that abound here in Florida Bay, it stabilizes the environment. Mangroves start by growing in open salt water and as the aerial roots spread, they gradually trap sediment and build the islands. Similarly, the sea grass, with its elaborate root system, keeps the soft, muddy substratum in place and prevents the waves from washing it away. The grass helps to construct the mud shoals on which it grows, and guides the course of the water currents in the entire area. It reduces the velocity of the currents and hence stimulates the settling of suspended particles, providing habitats for a great variety of organisms: not only *Melobesia,* but snails, clams, foraminifera, bryozoa, sea urchins, worms, fish, and many others. Sea grass alters the nature and distribution of sedimentary rocks.

Inorganic Precipitation?

Melobesia is not the only mud producer in the area. Attached to the leaves of the ubiquitous sea grass one finds huge numbers of serpulid worms, housed in their calcified tubes, and tiny shell-forming foraminifera. Elsewhere, free-living green algae contribute prodigiously to the accumulation of mud. Although Florida Bay is a luxuriant garden of chalk-mud producing organisms, some geologists believe that the shoals are mainly formed without the intervention of organisms. They note that throughout the summer, the water enclosed in these shallow lagoons is heated and evaporated by the intense sunshine. As a result, the concentration of salts in the water increases, so that they should be forced to precipitate. Under these conditions needle-shaped crystals of chalk, identical to those in the muds, should spontaneously crystallize out of the water. So, in their view, the mud is largely a physical and chemical product which precipitates directly out of the seawater as a fine suspension of particles.

It has been calculated, however, that the productive activity of the organisms in the bay could easily account for the total amount of chalk mud. In addition, there is reason to believe that spontaneous precipitation of calcium carbonate from these waters is strongly inhibited and cannot proceed at a massive scale without the help of organisms.

Precipitation Inhibitors

Some years ago, the late biochemistry professor Albert Lehninger said that there are such high concentrations of calcium and phosphate in our body fluids that we would follow the example of Lot's wife and literally petrify if there were no potent inhibitors in our systems to prevent us from turning into stone. Lehninger, whom I had invited to speak at a meeting on biomineralization in the Netherlands, went on to discuss some substances in our blood and other body fluids that inhibit the formation of crystals. Ironically, calcium phosphate formation is allowed to proceed in the human body, but only at special, dedicated sites—our bones and teeth—and under a strictly controlled biochemical regime. It is as if a window in the inhibitory system were deliberately created where crystals are needed.

In Florida Bay we are dealing with calcium carbonate, not phosphate, so is the analogy useful? Possibly it is. We have seen that, because of evaporation, the concentrations of calcium and carbonate in the water of the lagoon may become very high. Yet, spontaneous precipitation will not occur in significant amounts. Magnesium, which is abundant in seawater, strongly inhibits the precipitation of certain forms of calcium carbonate. And even more efficient are the slimes that make everything under water feel sticky. When I grabbed some of the mud and rubbed it between my fingers, it felt like yogurt. Everything—the mud, the grass, the seashells—seemed to be covered with a thin film of slime.

Slime consists largely of acidic polysaccharides, the kind of material that my lab has isolated from *Emiliania*'s coccoliths. That substance also interferes with the precipitation of calcium carbonate. In Florida Bay, slime is produced on a huge scale. To me, it seems as if the spontaneous formation of calcareous mud in the entire lagoon were rigorously held in check by living systems, as does calcite in *Emiliania* and calcium phosphate in the human body.

The chalk mud is produced at a tremendous rate, but only at specific sites, inside living organisms and under their control. And although in many cases the high concentration of salts in the ambient seawater may facilitate mud production, it is not the major controlling factor. Philip and I believe that chalk production in these waters is induced not by heat and evaporation, but by visible light emitted by the sun. Most of the chalk-producing organisms and communities depend on photosynthesis for their energy, just as plants do. With their green and red pigments they capture the sunlight, and this allows them to take carbon dioxide out of the water and to convert it into the organic matter of which they consist. As carbon dioxide is removed from the seawater, carbonate is produced, which associates with calcium to form a calcium carbonate precipitate. The calcifying organisms regulate and exploit the precipitate, confining it to specific microenvironments and using it to produce their skeletal structures.

This dependence on visible light and biological microenvironments, rather than heat and general high concentrations of calcium and carbonate, may have important consequences for the dynamics of chalk-mud formation. It suggests that instead of relying strictly on chemical precipitation, the system displays vast resilience and the potential for self-perpetuation. As soon as the ambient conditions permit mud-producing organisms to establish themselves, the whole machinery begins to operate. Chalk mud will be turned out in vast amounts, and this will form a suitable substrate for further colonization by the mud producers. Once the factory is switched on, it tends to reinforce its own operation and propagate itself until it reaches the limits of its domain.

Reef

The narrow string of islands extending from Miami to Key West comprises the Florida Keys, the emerged remnant of a fossil reef complex and some tidal sand shoals, all dating from the last couple of million years. The sheltered, shallow lagoon of Florida Bay lies behind them, while on the Atlantic side of the Keys one finds living communities even more varied than those in the lagoon. Algae—red, brown and green, many of them calcifying—come in all shapes and sizes. There is an astounding variety of sponges, mollusks, sea urchins, corals, and fish. And, as the sea floor starts to slope toward much deeper water, there lies the living reef itself. At the living reef the ocean gives the most spectacular display of its potential as a life-supporting environment. Huge colonies of corals coexist with extensive chalky crusts of red algae, shoals of brightly colored tropical fish, barracudas, and the occasional shark.

What a striking contrast to the lagoon. There we waded through soft, slimy muds; here, everything is hard, solid and brittle. The reef is a compact, rocky framework able to withstand the waves of the ocean even during the most devastating hurricane. In the lagoon, the enclosed water is often very salty. And although chemical precipitation is repressed by the slimes, there is some question as to what extent biology contributes to the formation of the muddy substratum. The reef itself, however, grows out into the open ocean, and its ramifying constructions are all made by living organisms.

The reef and the lagoon are similar in one respect. In both systems, the production of calcium carbonate is enormously stimulated by photosynthesis. The corals are made up of the calcareous skeletons of millions of minute animals, which form a living veneer over the coral and add to the mineral structure below by secreting calcium carbonate. While these animals are not themselves photosynthetic, in their tissues they nurture extensive colonies of small, photosynthesizing single-celled algae.

Reef-building corals are symbiotic: the animals and the algae are complementary in many ways and take advantage of each other's metabolism. The algae catch the solar energy, and with their excretory products they fuel the animals that secrete the calcium carbonate. The skeletons are produced only during the day; at night, the process comes to a standstill. This combination of photosynthesis and calcification is highly efficient. The rate of skeletal production is tremendous: a coral twig may grow 4 inches in a year. As a result, the reefs grow out explosively whenever conditions are favorable. This dynamic potential complements the balanced behavior of the community, allowing the living reef to rebound rapidly after storms, sea-level changes, and other potentially disruptive events.

Platforms and Atolls

Florida Bay is a carbonate platform. The whole area, including the reef, is made of calcium carbonate, or limestone, and is as flat as a pancake. Beyond the reef, the sea floor slopes steeply towards much deeper waters. Such platforms occur in the warmer parts of the world, in waters of the continental shelves where land-derived clays and sands have no access. Today, there are not many places on earth where you find them. In addition to Florida Bay there are the Bahamas, Mexico's Yucatan, the Persian Gulf, parts of the Sunda Strait, the Great Barrier Reef area, and a number of atolls in the Pacific. There were times when they were widespread and much larger. In other periods they were almost absent. Platforms have waxed and waned over geological time.

The Florida Bay platform is an example where incredible amounts of limestone have been produced over the ages. You have to drill for more than 3000 feet before you reach a different rock type, and that is of Cretaceous age—some 70 million years old. Not only that, but the platform sheds most of its products into the surrounding ocean waters so that enormous carpets of limestone are accumulated at its feet in the deep sea. Here is an interesting problem: A pile of shallow-water carbonates is 3000 feet thick. Every foot of the pile was deposited near sea level, within reach of the sunshine, but the oldest parts are now deeply submerged. We know the sea has not risen 3000 feet, so the platform must have subsided. How has the top of the platform always stayed just near the water surface?

This platform has grown in such a manner that suitable conditions for its survival and proliferation are maintained. It keeps its crest just beneath the water surface, in spite of subsidence. The chalk-secreting communities at the surface are the most productive in that environment, and so they automatically preserve optimum exposure to the sunlight and perpetuate the entire structure. Another way to look at a platform is to regard it as a solar collector of immense sophistication and size built by life from the floor of the ocean, a self-sustaining biological tissue covering its flat, submerged top.

Although they are relatively small, the most elegant carbonate platforms are the atolls (fig. 7.2). These start off as fringing reefs growing up in the shallow water surrounding a volcano in the tropical ocean. The volcano slowly sinks, either into the ocean crust under its own weight, or because plate tectonics pushes it away from a mid-oceanic ridge. But the reef continues to grow, even when the top of the volcano subsides deep below the water surface. It forms a circular ring surrounding a shallow lagoon. As around Florida Bay, the reef sheds its limestone debris along its sides. Eventually, an enormous, flat-topped pile of limestone, up to 3000 feet thick, accumulates on the top of the volcano. When this construction is pushed out of the tropical zone by the moving plate, the platform finally dies and continues its path as a flat-topped submarine mountain.

These model carbonate platforms are beautifully functional. A cliff at the shore is undermined, its supporting rocks erased by the waves. But the reef grows out into the waves, thriving in turbulent regimes. It forms a protective

7.2 An atoll: a solar collector built by the biota atop a drowned volcano.

ring around the lagoon, and grows upwards while the volcano sinks into the deep, so that its crest stays level with the water surface. The lagoon provides internal support for the upright tube of limestone produced by the ring reef. (Again, the rock-building living community there automatically maintains its strategic position in the upper, sunlit domains of the water where it can take full advantage of photosynthesis.) Finally, the debris around the outer slope of the atoll is hardened and cemented together, providing a stable base for the ever-rising platform.

Open ocean water in which atolls grow is notoriously poor in essential nutrients such as phosphate and nitrate. These lonely platforms feed themselves by efficiently scavenging any of these substances from the water washing over and percolating through the atoll. Once the nutrients are captured by the organisms they may be recycled again and again. The submerged volcano may be an additional nutrient source. The gigantic calcium carbonate structure may act like the wick of an oil lamp: evaporation at the atoll's surface may set into motion a current of water up through the volcanic cone, leaching the nutrients out of the basalt and bringing them to the covering film of biota. One may compare an atoll with a beautiful oasis maintaining itself in a liquid desert.

Calcium Carbonate and Evolution

Carbonate platforms are remarkable biological systems. By studying their geology, we can reconstruct their evolution. It is useful to look at the process of limestone production from the vantage of the rock cycle, and to place it in an historical perspective. Limestone on the continents is transported, mostly in solution, by rivers flowing toward the sea.

Assuming that the oceanic composition is more or less constant through time, the dissolved calcium and carbonate cannot remain there: they have to be removed at about the same rate as they enter. So we can think of the carbonate platforms as oceanic sinks of calcium and carbonate. Eventually, these big bodies of limestone are plastered back onto the continents by plate tectonics, and the cycle starts again.

It is likely that early in earth history, limestone was removed from the oceans inorganically by solar heat, through evaporation from restricted basins. But there is evidence that true carbonate platforms were already in existence 2.5 billion years ago. In those days, enormous carbonate accumulations were made by the bacterial communities that form microbial mats. One can find their fossilized remains, laminated carbonate rocks called stromatolites, in the central parts of all the continents. Like the present platforms, stromatolites were probably energized by visible light. Photosynthetic bacteria were the dominant organisms.

Then, at the beginning of the Cambrian Period (around 570 million years ago), there was an outburst of new calcified forms. Suddenly, a huge variety of animals began to produce shells with a precisely defined structure and shape. From then on, the carbonate platforms were no longer dominated by bacteria, but by animals and algae.

This development provoked a revolution in the calcium- and carbonate-removing capability of the oceans. Present-day stromatolites grow very slowly, and usually the corals and algae that make up the platforms easily outpace them. The new platforms must have been able to respond more flexibly to sea-level changes and to variations in the calcium and carbonate fluxes from the continents into the oceans.

Yet another new element was introduced at this point.

Shell-forming animals had brought calcification under control to such an extent that in many cases they could uncouple its operations from photosynthesis. Large-scale limestone production could spread to higher latitudes, far outside the tropics. That may have provided means for a considerable geographical extension of the oceanic calcium carbonate sink. All as a result of enhanced biochemical control of calcification.

As the sink was brought more and more under control by the biota, its operation was progressively freed from simple physical and chemical constraints. A further advance occurred some 200 million years ago, with the advent of the coccoliths and other calcifying plankton. They carried light-stimulated limestone production out into the open ocean, away from the shelf seas. Large parts of the ocean floor were added to the sink of calcium and carbonate; no longer did the removal of these materials from the seas depend on suitable conditions for platform development in the peripheral zones of the oceans. The waxing and waning through time of the carbonate platforms could now be absorbed by fluctuations in the production of calcified plankton.

Limestone is a major reservoir in the earth's crust that, together with the organisms responsible for its formation, has evolved through time. This evolution is tied up with the history of the oceans and of the earth as a whole. Although we are not sufficiently informed about the biochemical mechanisms underlying the production of this reservoir, we can now discern some general principles apparent in *Emiliania* and carbonate platforms alike. The bulk of calcium carbonate deposition is tied up with photosynthesis rather than with high concentrations of calcium and carbonate in the main bodies of ocean water. Instead of heat radiation, the visible part of the spectrum of sunlight is the main driving force. Inorganic precipitation is suppressed by polysaccharides and other slimy materials. In minute pockets, the biota create a broth in which crystallization is allowed to proceed in an organized fashion. That is the factory of limestone: a rock that evolved.

8

HUMBLE FOUNDATIONS

When confronted with slime and bacteria we instinctively recoil in disgust. These phenomena evoke visions of disease and decay, and are similarly repulsive and untouchable. In like fashion, we impatiently scrape mud from our boots and bemoan the rains that transform our manicured lawns into soggy mush. How little do we realize that slime is beautiful. Or that bacteria play a key role in the regulation of all the fluxes of matter through the biosphere. Or that the lowly mud beneath our feet covers much of the earth's surface and is a universal substrate for life. Such are the humble foundations of the living world.

Slime

Planets like Mars and Venus have a loose, rocky covering, whereas the earth has muds, mats, soils, peats, and sedimentary rocks. Much of the difference between Mars and Venus on one hand and the earth on the other is due to slime. Slime is the glue that holds the biosphere together. In bacterial and diatom mats this substance keeps the sands and clays of shores and soils in place. It helps trap sedimentary particles and build islands such as Vlieland. Mussels and cockles wrap suspended sedimentary particles in their slime and force them to settle on the ground even in turbulent waters. Slime inhibits limestone precipitation in Florida Bay and it may be involved in the multiple transformations of that sediment after it has been formed.

Thus, it seems, slime is multitalented. Even the human body cannot function without it. In an organism that harbors the intellect that shuns it, slime forms a thin veneer covering the surface of the digestive, respiratory, urinary, and sexual tracts. In this strategic position between the invaginations of the outside world into our bodies and the sensitive tissue lining, it performs an infinite variety of mediatory functions. Slime harbors antibodies and iron-capturing proteins and so forms an impenetrable barrier to infectious microbes. In the digestive tract, the layer of slime we call mucus prevents any bulk flow of molecules into and out of the tissues. It reduces the mode of transport of these materials to one of pure diffusion, attenuating the rate of their delivery and thus providing time for the cells to deal with them in the most appropriate way. If the tissues lining the stomach were not protected by this specially tailored buffering layer, they would be digested in the aggressive, acidic environment they envelop. Slime is intimately mixed with our food on its long passage through the gut. The resulting paste is easily transported, digested, and excreted.

If you are not too much of a smoker, a constant stream of particles, captured from air entering your lungs and subsequently wrapped in slime, moves upwards along the walls of your trachea to the throat, from where it is continuously swallowed and transported to your stomach. Through this circuitous route, the digestive tract is cleansed. Slime lubricates your joints, and also your tongue: in its absence you wouldn't be able to speak.

An immense variety of long molecules falls under the collective name "slime." As a rule, the main constituent of slime is a carbohydrate carrying a negative charge. Many such carbohydrate chains may adhere to a single protein molecule to make a complex that looks like a brush. Large three-dimensional networks may be formed when the brushes are linked up, for instance by atoms with two positive charges, such as calcium. The more numerous the links between the brushes, the more syrupy the constitution of the slime. However, once you shear that sticky material and deform it past a certain point, the molecules pull apart from each other and the slime becomes liquid. A slug makes good use of this property. It rests one part of its foot on the ground, allowing the slime molecules to entangle; the animal is now firmly anchored. It then drags the other part of the foot over the surface, forcing the slime to assume its liquid and lubricating properties. Gliding here and adhering there, and then the other way round—this is the slug's mode of locomotion.

The macromolecular part of slime is only a loose skeleton holding the structure more or less in shape. Macromolecules make up less than one percent of the slime; the rest is water. So, it requires little investment to make a lot of it.

Furthermore, nature is thrifty: most organisms either eat and recycle most of the slime they produce, or use other organisms to mediate in the recycling process. For example, mollusks like limpets and chitons (slug-like animals that live on rocky shores, feeding on algae and bacteria) leave behind a slimy trail that absorbs nitrate and phosphate from the surrounding seawater. Microbes thrive on this trail, photosynthesizing to form a luxuriant garden that the animal harvests on its return to the scene. Brachiopods and other organisms likewise shed mucus in great abundance to cleanse their tissues.

The structure of slimes is meticulously tailored to fit the particular function required. The viscosity may be regulated not only by shear, as in slug locomotion, but also by subtle chemical modifications at specific sites. Such changes regulate a woman's fecundity: during short intervals in her monthly cycle, the secretions in her sexual tract become sufficiently fluid to allow sperm cells to reach the ovum.

Slime is indispensable even on the small scale of the human cell, each of which is covered with a thin slimy coat. This veneer plays a unique signaling function: the whole biochemical machinery of a cell may change when its slime cover is recognized by that of another cell. Such an event may influence whether the cell becomes part of the liver or the skin, for instance. Thus, slime may mediate the diversification of tissues and the emergence of an animal from a fertilized egg. Slime is an ugly duckling. The more we know about it, the more it seems a beautiful swan.

Symbiosis and Evolution

Not long ago, it was my practice to teach my students that evolution is progress from the simple to the more advanced and complicated organisms. I implicitly regarded humans as the ultimate crown of nature. I ignored bacteria, as did the textbooks at my disposal. I restricted my lectures to what I thought were "higher organisms," and believed that the major division among living things was that between the plants and animals. Lynn Margulis, a professor in the Botany Department at the University of Massachusetts at Amherst, convinced me otherwise by reminding me of our bacterial beginnings and humble place in nature.

Emiliania, the unicellular alga with its coat of calcium carbonate, helps illustrate Lynn's argument. With an electron microscope we can glimpse how an *Emiliania* cell is organized (fig. 6.2). In the center there is a nucleus, a little bag enveloping the genetic material, DNA; there are one or two chloroplasts, little green bodies specially equipped to carry out photosynthesis; a mitochondrion, designed to catalyze the respiratory reactions; and finally several other membrane-bound pockets, or vesicles, including the one in which the coccoliths form. Bacteria, on the other hand, have no nucleus, and few organelles and compartments where special functions are carried out for the cell. Their biochemical machinery is just a tiny blob of protoplasm enveloped in a bag of lipid membrane, cell wall, and mucus. This is a major division among living things, more fundamental than the difference between animals and plants: on the one hand the prokaryotes, or cells without a nucleus, and on the other the eukaryotes, or nucleated cells, to which *Emiliania* and we ourselves belong.

Lynn is an outspoken champion of the theory that the eukaryotic cells evolved some 1500 million years ago through the physical association of different bacterial, prokaryotic, ancestors. This arrangement in which partners live together throughout their life cycle is called symbiosis. Lynn's message is startling for those who cherish the conventional wisdom. We used to believe that evolution implied continuous struggle for survival among species and individuals. Largely through Lynn's lectures, books, and films, however, the idea of symbiosis in nature has gained wide acceptance in the scientific community. She regards evolution from a prokaryotic point of view, emphasizing the fundamental importance of bacteria for the earth at large.

Bacteria were the earths' first inhabitants. Their remains are found in ancient, flinty rocks on all the continents; the oldest found so far, 3.5 billion years of age, were discovered in South Africa and Warrawoona, Western Australia. For 2 billion years of this period, they were the earth's sole inhabitants and used this time to develop all of the major cellular metabolic pathways found in bacteria today. In the bacterial world, one finds virtually the entire photosynthetic apparatus that the rest of life uses to capture its energy. Collectively, the bacteria have the catalytic potential to make use of all the chemical elements needed for life; of extreme temperatures, pressures, and elevated salt concentrations; and most of the more moderate conditions available in the biosphere.

Bacteria never work alone; they always collaborate. A single bacterial type can conduct only a very limited repertoire of reactions, but the waste products from these—gases and solutions—accumulate in the environment, providing an opportunity for other organisms with complementary properties to proliferate by using the garbage. In this way a complex network of interacting bacterial types is eventually established, conducting the nutrients along circular pathways and recycling these valuable substances over and over. Poisons are diligently kept away from these cycles. And when conditions change, new teams automatically move in to ensure the further proliferation of life. From what secret reservoirs do all these microbes stem? Why are they there, ready to take over whenever new conditions are created in the evolving biosphere?

Lynn studies microbial mats that occur in salty, sheltered lagoons off the Pacific coast in Baja California, Mexico. These mats, densely populated by bacteria, grow abundantly. There are a huge number of mat types, each representing a different community of myriad interacting bacteria. Lynn concentrates on just one of these communities, dominated by the green, photosynthetic cyanobacterium *Microcoleus.* Lynn and her students have found that each cubic inch of these mats contains perhaps as many as two hundred different types of microbes, mostly bacteria, but protists and fungi as well. At any given moment most of these species are dormant, barely participating in the metabolism of the bacterial tissue. After several years of study, a flood engulfed the mats, and for two years the area was transformed into a freshwater lake. Then, salty intertidal conditions slowly returned. The researchers followed the accompanying changes in microbial life and saw that each alteration brought different microbial teams into action. It became obvious that the vast diversity of microbial life in the mats acts as a reservoir for potential activity, with dormant species waiting to take over when conditions favor them. Lynn suggests that the mats of *Microcoleus* are analogous to the entire biota which inhabits a vast variety of terrains, from the origins of life up to the present day.

Another, related aspect of bacterial dynamics is crucial to Lynn's view of the importance of microbial interaction in earth history. Canadian microbiologists Sorin Sonea and the late Maurice Panisset were intrigued by the fact that in bacteria genetic information can pass from cell to cell, and from one type to another. Unlike the rest of the living world, bacteria are not restricted to giving their DNA to offspring; they can also pass it on to their neighbors. These transfers of DNA are common, which means, in geological terms, that new inventions can rapidly permeate the bacterial world. Sonea and Panisset suggested that the entire bacterial community could be considered as one earth-embracing organism, drawing on a single gene pool, infinitely rich in its catalytic potential, continuously reorganizing itself, adapting to the ever-changing conditions in the biosphere. This view is sweeping in its simplicity, but I am inclined to believe that there is some truth in it.

The effects of all this activity are astounding. The microbes cleanse the biosphere and keep it fit and nutritious for life, as we shall see in the following chapters. Each organism, just by living, adds chemicals to the environment, the waters, surface sediments, and atmosphere, and removes others. Carbon dioxide, oxygen, methane, hydrogen sulfide, and nitrogen are taken up, transformed, and then returned, on a worldwide scale. Bacteria play a prominent role in the chemical household of the planet; they are indispensable for the provisioning of the rest of life.

Eukaryotes

How fruitful was the association that led to the eukaryotic cell! The economies of symbiosis unleashed an avalanche of startling innovations. Not at the metabolic level—virtually all the metabolic reactions of the eukaryotes had already emerged in their prokaryotic ancestors. Indeed, they constitute only a limited selection of the gigantic range of reactions that bacteria can catalyze. No, eukaryotes are specialists. Their evolution has been one of regulation, organization, differentiation, size, and complexity. *Emiliania* is a case in point. Many bacteria can induce calcium carbonate formation, but none can make anything as complex as a coccolith. And this is just one example out of thousands. Nature abounds with single-celled eukaryotes. Amoebas, paramecia, foraminifera, heliozoa, and diatoms are just a few of the better-known examples.

In some eukaryotes, known as animals and plants, the cells have acquired the potential to arrange themselves in elaborate arrays. In contrast to most communities of their bacterial ancestors, these eukaryote colonies are not associations of different organisms selected from the inexhaustible pool of genetic variation available in the environment. All individual cells contain precisely the same genes; they are a single family, derived from one precursor: the fertilized egg. Yet, they can specialize and carry out a wide variety of functions. Complex mechanisms have emerged that inhibit the full deployment of the common gene pool in the individuals. In each type of cell, each tissue, only a limited set of genes is allowed to be expressed. The growth of an animal or plant proceeds through a process of cellular differentiation. It is a masterpiece of self-organization. Minor flaws at early stages may cause the organism to die; at least they will threaten its survival.

Despite their inventiveness, the eukaryotes have a metabolic range so narrow that they cannot recycle most of their nutrients. Bacterial support is essential for survival. Nor do eukaryotes have a universal pool of genes at their disposal; large genetic innovations cannot readily permeate their world. The very complexity of organization largely precludes "horizontal" passage of DNA from one individual to the next. Genes are transmitted "vertically" from parents to offspring.

Animals and plants are newcomers on the evolutionary scene. The earliest reported animal remains are 700 million years old, a mere fifth of the age of the earliest known bacteria. The familiar types of plants and trees came even later—some 500 million years ago. Bacteria, with all their diversity, have been a resounding evolutionary success. No major metabolic type has apparently gone extinct, and even today bacteria dominate the natural environment. However, the fossil record reveals that eukaryote species live in a house of cards in which any given species is vulnerable to any major disturbance of the global environment. The past 600 million years have been punctuated by catastrophe upon catastrophe, each of which wiped out large sectors of eukaryotic life. In contrast, the history of life in the Precambrian—the period of geology before animals with skeletons came into being—appears to have been one of steadily increasing complexity, despite a few major cataclysms. The present environmental change that results from human intervention may be a threat to our existence, but to the bacteria it can be no more than a ripple on the surface.

Mud Mounds

Once you know what to look for, you find slime everywhere. It is the matrix of the bacterial tissue in microbial mats. It is the substrate for "marine snow," the microbial microcosms that drift as fluffy flakes in the oceans. It binds heavy metals and keeps them away from the microbes. It pervades our soils and mud flats. It collects sediment particles and forces them to settle. It covers deserts and dunes, helping to capture and retain water.

It is little surprise that with their multiple talents, slime-building bacteria prove to be major players in certain geological dramas. Nowhere, perhaps, is this more striking than in the production of carbonate rocks, most notably those like the stately dome-shaped limestone formation near the town of Couvin in southern Belgium.

One reaches this curious formation by driving through the Ardennes, going south from Liège. The road follows meandering rivers cut into an elevated plateau. Much of the original land has vanished due to intense digging, carving, and quarrying of the ubiquitous marbles and limestones. Today, however, that activity has largely ceased because of competition from outside Belgium, and the quarries are abandoned. Rusty cranes, crumpled conveyor belts, and scattered blocks of limestone bear witness to fairly recent activity, but the deepest pits are filled with water, and the steep walls of the quarry are overgrown with moss and shrubs. Little by little the old quarries are being reclaimed by nature, playing out the age-old struggle between humans and rocks.

If you follow an ancient path carved through the woods, you reach an open space dominated at the far end by a vertical wall of rock 600 feet wide and 150 feet high. It is a section through a dome-shaped limestone body, neatly carved as if with a razor blade. Reddish-gray marble, as pure as can be, rests sandwiched between layers of clay. The rock is lonely and huge, a local anomaly enveloped in a mass of hardened, land-derived debris. From a distance, it is clear that the limestone body consists of slabs piled one atop the other and lying parallel to the upper surface of the dome. At the sides of the dome, these slabs lay at an angle of 45 to 50 degrees.

Closer to the wall, one sees colonies of fossil corals, both flattened sheets and branching limbs. There is no doubt, these rocks must have formed at sea. But curiously, nowhere had the corals built a solid framework. It is quite unlike reefs, where the corals are connected, each growing from the surface of an older skeleton to form a structure that can resist the ocean waves. Instead these colonies are disconnected, surrounded by a matrix of limestone, dense and red in some places, fine-grained and gray in others.

There are many similar limestone edifices in these mountains. With their dense limestone matrix and their strange content of corals and other fossils they have long captured the imaginations of both geologists and passersby. For years this limestone body was viewed as an ordinary coral reef. Geologists assumed that the matrix had been formed by coral skeletons that were destroyed by recrystallization long after the limestone was buried in the clays. But this did not explain why some of the corals are so well preserved. Other geologists proposed that the structure was a heap of lime debris, swept together by waves and ocean currents. But in that case, why was it not mixed up with the clay all around?

My friend the Belgian geologist Claude Monty finally reconstructed the peculiar geological events that built the dome. He came to the area with the experience fresh in his mind: he had studied microbial mats in the Bahamas, and spent many months on Australia's Great Barrier Reef. Instead of seeing only the bare sedimentary floor, as his traditional geological training encouraged, he realized that the whole Recent environment was living and crawling, transferring material, and producing sediments, a frenzy of activity he calls a "factory ruled by life."

It was not at the reefs themselves that Claude became aware of the important role that bacteria play in the production of these carbonate rocks, but in the more protected lagoons behind them, where large accumulations of white, soft, and slimy chalk muds cover vast areas of the shallow, sunlit sea floor. If he scooped off the top layer of mud and studied it before it dried, he found slimy bacteria on a massive scale.

With this example in mind, one can look at the fossil limestone in the quarry in a different light. Many of the lopsided limestone slabs, for instance, are intensely distorted. They had once behaved as a cohesive, gelatinous paste. The liquefied material had been compressed by the weight of the overlying rocks, and was then squeezed and injected into them. There are injections all over, or at least their solidified remains.

"We now believe that this entire dome was a mound of slime, bacteria, and limy mud," Claude says. "When it was formed, it was a cohesive mass, and this explains the steep slopes along the flanks. An earthquake or some other disturbance destabilized this pile of sludge and triggered off internal distortions, collapses, and injections." From a distance of some 50 yards, one notices a couple of lopsided slabs inside the carbonate body near the top of the hill. They are blocks of limestone, leaning against each other like a Gothic vault. This was the result of a collapse. The area below the arch had been empty at first, but then filled with limestone, arranged in horizontal laminations. These horizontal fillings are stromatolites, the hardened remains of microbial mats that inhabited the cavities shortly after the collapse took place and when most of the mound was still in its gelatinous state.

"You can still see the remains of the bacteria in these rocks, and in the cavity fillings," Claude says. "The main slabs were big stromatolitic structures, but their internal structure was distorted by the traffic of animals through the sediment surface. And the corals? They were sitting on top of the mound until the substrate moved. They fell aside, subsided, and were covered by sediment."

The mounds are about 350 million years old. In geological terms, they were formed quickly, in no more than 1.5 million years. During this relatively short period, there must have been a halt in the normal sedimentation of clay, which gave the limy mound a chance to form. Soon after, it solidified into rock. The overlying clays did not penetrate the structure, although there are many ancient cracks and fissures. These must have been filled with lime and hardened before the clay reappeared on the scene.

Claude and his colleagues have found many mud mounds, both fossilized and recent. They occur along fault zones in the deep sea, far deeper than the shallow waters where reefs can grow, and develop during periods when the sedimentation of land-derived debris is temporarily interrupted. Claude believes that they are fed by organic materials washed in along the continental slope or brought there by deep ocean currents. Maybe the bacteria that build the mounds also exploit local seepages of nutritious gases, such as methane, expelled from the surrounding rocks and emanating through the faults along which the mounds occur. Specialized bacteria have been described that, using oxygen, can feed on such gaseous exhalations in the deep sea, and that, in turn, support elaborate living communities.

Mud mounds are gigantic phlegms of nature, quivering masses of bacteria, slime, and minerals. How curious that they do not arouse our disgust, but our admiration. Instead of averting our eyes, we unearth them with care, slice and polish them, and use them to adorn our banks and other prestigious buildings. No less respect is due our bacterial partners that live within us and make our lives possible, in a symbiotic arrangement of which we are only passingly aware. We, too, are made of bacteria, and are held in shape by slime and biominerals. Some mud mounds will endure, however; we are more transient features.

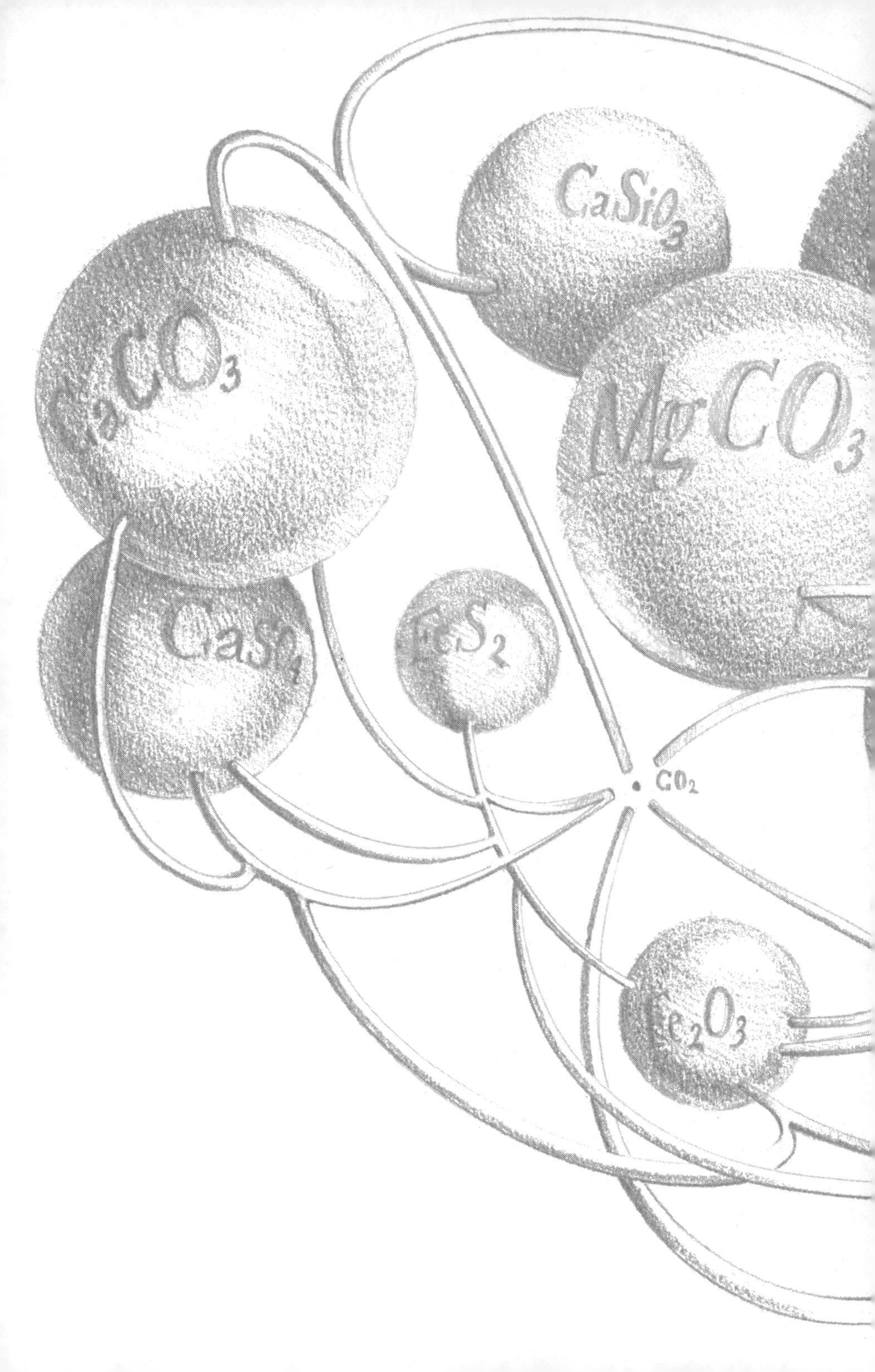
$CaSiO_3$
$MgCO_3$
CO_2

PART III

Life at the Planetary Scale

9

BIOSPHERES

IN THE LEFT CORNER OF my study window a glass sphere about eight inches wide rests on a simple plastic stand. It looks like a fishbowl three-quarters filled with water, with some fine gravel at the bottom and a miserable-looking twig with a greenish cover of algae. Ten shrimps, each about a third of an inch long, swim nervously when the bowl is disturbed.

This is no ordinary fishbowl. It is a closed vessel, totally sealed off from the outside world by a spherical glass wall. The organisms inside live in a world different from ours. They have no access to the air you and I breathe. Nothing can enter or leave the interior. There is no way to feed the shrimps, or to refresh the water. Although sunlight sustains its only energy, no matter enters or leaves. It is a totally self-regenerating system, a little cosmos, budded off from the world at large. It operates according to its own rules and laws, largely independent of our familiar world. Spheres of this kind contain the only "extraterrestrial" life known to humans.

Shrimp and algal life inside the sphere has gone on for more than two years now. According to EcoSphere Associates, the company in Tucson, Arizona, that supplies these desktop biospheres, the shrimps in my glass bowl may survive another ten years or so. Ten years! The first closed ecosystems, without shrimps, were started back in 1967—and they are still alive. When I first saw one of these spherical aquaria I felt like a child who for the first time recognizes himself in a mirror. While looking at the sphere I became strangely aware of myself as an object confined, like the shrimps, to that other materially closed entity: the earth. We do not live *on* the earth, as we tend to believe, but are engulfed inside it. As the magician and philosopher David Abram has said, we move around inside an invisible, yet dense and tactile shell of the earth, the atmosphere. We feed on this gaseous medium, and dump much of our waste into it. We transform it, and yet we take for granted the unlimited supply of this salubrious broth.

On our planet, life has gone on vastly longer than in the oldest of the artificially confined biospheres, and it has lasted for 3.5 billion years without interruption. Whatever wholesome principle keeps the earth alive is in this crystal, it seems, split off, isolated like a purified protein in a test tube. What, then, is the secret? What keeps the water clean? Why do the shrimps not smother in their own excrement? What keeps the algae from rotting away? What powers keep this microcosm in balance?

The National Aeronautics and Space Administration is behind this contraption, as part of its Controlled Ecological Life Support Systems (CELSS) research program. Prolonged trips into space require inexhaustible supplies of food and air, so the scientists at NASA set about designing a self-regenerating system that can support entire crews, in principle indefinitely. The sphere on my window sill is a simple classroom prototype; of course, much larger systems will be required for space. A major objective or research on such closed ecosystems is the colonization of Mars. Eventually, elaborate installations should be sent there to make that planet habitable.

Independent of NASA, Space Biospheres Ventures, Inc., in Oracle, Arizona (near Tucson), is experimenting with what it calls Biosphere II—as opposed to Biosphere I, the one in which we live. Biosphere II is a huge terrarium with a bit of ocean, jungle, desert, savanna, and salt marsh, all on a miniature scale. Some hope to seal it off from Biosphere I for a hundred years. The first crew of eight people will agree to be locked in for two years. Airlocks and space suits will prevent contamination when crews are changed.

In addition to helping to move along the colonization of Mars, contraptions such as Biosphere II will be used as models by which to study the behavior of our own biosphere. *Biosphere* is the name for the sum total of all living things on earth and their environment. As such it is a purely descriptive term, not referring to any properties this collective may have over and above those of its constituent parts. Actually we are pretty ignorant about life at the planetary scale. Indeed, the enormous age and complexity and the gigantic dimensions of our biosphere as a whole almost prohibit any comprehensive study. By contrast, closed ecosystems are amenable to experimentation. Can they indeed be taken as reliable analogs of the real biosphere? Although we may feel that the two types of biosphere are closely akin, they may be entirely different in fact. The terms Biosphere I and Biosphere II are very suggestive, and they call for a critical comparison. A close look into the crystal sphere on my windowsill may be of some use for a start.

Cleansing

You don't have to look into the crystal sphere for long to get at least a basic idea of what is going on. To begin with, the sphere is only closed so far as matter is concerned; energy—heat and light—is readily exchanged with the surroundings. Sitting in my office, the baby biosphere is kept at a comfortable temperature; heat has no trouble getting in or out. Most significantly, the glassy enclosure allows the sunshine to come in. Ultimately, it is light that energizes the system, animates the organisms, and supports the activities of the shrimps, algae, and bacteria that clean the water and air.

The organisms in the sphere grasp the sunlight and use it to recycle the available nutrients and to keep toxic substances out of their way. The longevity of the crystal sphere demonstrates the extreme efficiency with which life can cleanse its own environment. If only a small fraction of any nutrient leaked away from the circulatory network, if any one toxic material had the slightest chance to accumulate where it could exert its deleterious effects, then the whole system would soon clog up and come to a standstill, or else change its behavior and composition entirely. Unlike the living things on the earth at large, these encapsulated organisms cannot depend on geochemical fluxes for their ultimate nutrition and waste disposal. True, the shrimps and algae are insignificant in comparison with the volumes of water and air enclosed in the glassy shell; nevertheless, this medium is all they have, both for sewage and as a nutrient source. Those who tend not to keep this environment clean and wholesome are selected against. Only those that help subject the sphere's interior to the demands of their entire organization can survive.

Algae and shrimps—how thoughtful the inventors were to place such fragile and elaborate organisms in this biosphere. Of course, the system would work just as well (or even better), with bacteria alone, but a complex community of shrimps, visible to the eye, keeping itself alive for five years or more illustrates wonderfully the cleansing powers of life.

What do shrimps need in order to survive for so long? Like us, they depend on organic materials for their subsistence. The animals ingest food, which, in their cells, reacts with oxygen retrieved from the water. Then they use the energy liberated during that process. The reaction products are the organic compounds of which the shrimps themselves are composed, and also carbon dioxide and water which are released as waste. In the confinement of their world, the creatures can survive only if the refuse is reconverted into food. It is here that the algae come in. With their green pigments, they capture some of the light entering the sphere, and in the act of photosynthesis they reconvert the carbon dioxide and water into organics and oxygen. This recycling of carbon by the two complementary systems, the shrimps and the photosynthesizers, is illustrated in figure 9.1. You could say that the algae eat products of the shrimps and vice versa. These components of the community act as coupled chemical reactors, driven by the energy of light.

Complexity

In reality, the system is a great deal more complex. For instance, in addition to carbon dioxide and water, the shrimps produce excrement that the algae cannot readily absorb. There must be armies of bacteria and protists in the bowl that feed on the less manageable remains and break them down into carbon dioxide and water. The shrimps, in turn, eat the protists or the bacteria, as well as waste matter the photosynthesizers undoubtedly excrete into the ambient medium. To further complicate the picture, the microbes are of many different kinds, each feeding on a different selection of the enormous variety of organic substances in the sphere. They are organized into teams of specialists and form an intricate network of diverse organisms with complementary tasks. The entire community works with the perfection of a sea gull gliding in the sky. Any perturbation is counteracted by an appropriate response of the whole.

Clearly, the simple representation of the organic carbon cycle given in figure 9.1 is inadequate. In fact, more than twenty chemical elements are needed for the maintenance of life. The most important nutrients are hydrogen, oxygen, carbon, nitrogen, sulfur, and phosphorus, but a whole range of other elements are required, including sodium, potassium, calcium, magnesium, copper, silver, zinc, and molybdenum. It follows that all these different materials must be cycled through smoothly operating, biologically catalyzed networks, just as is the case for organic carbon.

There are other complications that the living things in the sphere seem to overcome with ease. Not only do they keep elements available in the appropriate chemical form, but they actively maintain them at close to optimum concentrations in the water. If the concentration of a critical component is lower than that optimum level, the operation of the entire system will be curtailed; if it is higher, the material may no longer act as a nutrient but as a poison. Most elements are poisonous at high concentrations, and many others are toxic at any significant concentration. Examples of this latter category are elements that seem to have no biological function, such as arsenic, cadmium, mercury, lead, and uranium. The biological collective in our glass sphere not only collaborates to maintain an optimum distribution of all the nutrients, but it also has to keep the multitude of poisons and toxins out of the way. The organisms in the sphere in my office are involved in a cleansing operation so

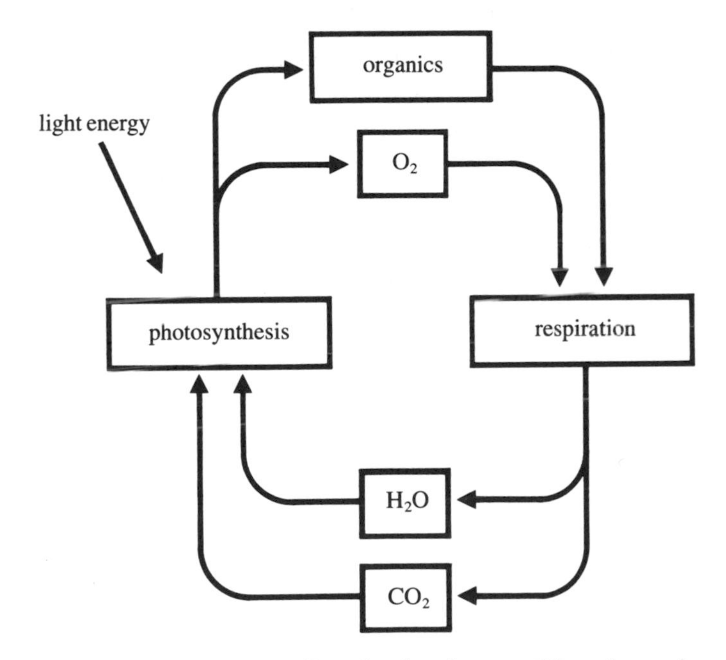

9.1 "The flywheel of life": cycling of carbon between CO_2 and organics, catalyzed by photosynthesis (the algae in the desktop biosphere) and respiration (the shrimps).

subtle, complex, and all-embracing that it defies description.

Leaving the sphere in the shade for a few days has a remarkable effect. The algae grow rapidly and even spread out as a thin, green film over the inner surface of the glass. It appears that the system rapidly adapts to the new conditions: the decreased exposure to light brings about an increase in photosynthetic potential. One may expect the other metabolic activities of the algae, the shrimps, the protists, and the bacteria to be accommodated accordingly, so that the entire machinery keeps running efficiently.

All organisms have an astounding potential for growth and multiplication. Some bacteria, for example, divide every twenty minutes as long as conditions are favorable. If a single bacterial cell were to multiply at this rate during only three days, its offspring would equal the earth in volume. This tendency towards explosive multiplication is strictly curtailed, both in nature at large and in the crystal sphere in my window. There must be heavy competition for foodstuff and energy among the organisms in the sphere. A process of natural selection is at work which evidently leads to the maintenance of a collaborating team perfectly capable of exploiting the conditions provided.

Cosmic Dimensions

Pondering the intricacies of nature is a superb form of entertainment basic to all scientific enquiry. It brings a sense of cosmic dimension to even the most insignificant components of the real world. The crystal sphere brings home the amazing complexity of life. It also reminds me of those Russian dolls that my children used to play with. You open one up, and out comes another. That one, in turn, contains an even smaller doll, and so on. Similarly, each living component in my crystal biosphere is a microcosm consisting of microcosms of even smaller dimension.

Figure 9.2 gives a simple outline of what goes on in a single cell. It includes some of the major metabolic routes along which foodstuffs are channeled and the metabolic interactions that occur in most organisms. There are many sidetracks and variations on this theme. The figure does not show how the genetic content of a cell is reproduced and brought to expression, and it ignores most of the mechanisms that regulate the cellular activities. Nonetheless, the figure suggests that the living cell is an infinitely complex chemical factory.

In the figure, one sees a multitude of linear as well as cyclic routes. At the periphery of the network are several open-ended pathways: nutrients are captured and conducted into the system on one side, and at the other side they are released as wastes. In between, the chemical processing proceeds with incredible smoothness.

When you open up the cell and look inside, you see that each of the nodal points on the maze represents a particular chemical, and each connecting line a particular enzyme, a member of the macromolecular world. An enzyme is a microcosm unto itself. It has the specific function of speeding up one particular reaction, in the metabolic web. In figure 9.1 the simple expression "organics" refers to organic matter, and the subtle world of biochemistry, of proteins, nucleic acids, carbohydrates, lipids, and metabolic intermediates with their myriad forms and interactions. Each macromolecule is finely tailored to perform a specific function in the cell.

Our crystal sphere is a bewildering world. Each cell type, each biological species, is idiosyncratic, with its own preferences, needs, and dislikes. What is waste for one organism is food for another. The different metabolic networks in the bowl must be complementary to a remarkable degree. They align themselves in the fluid medium to form the master maze through which the nutrients are always returned to their origin, and by which the toxins are kept out of the way. In fact, reality is infinitely more subtle than this. The whole system is constantly renewing and reorganizing itself, and all we have imagined so far is a transitory moment in a long sequence of events—or rather, a mean steady state maintained only until the system assumes a new configuration.

Each component of the web, each protein even, has a limited life span. It is destroyed after a while, conducted through the recycling network, and replaced by a new specimen. When conditions are favorable for one of the organisms it will make a copy of itself, and will continue to do so until the environment becomes less advantageous. Other organisms may be tuned down when there is no place for them in the sphere. They die, or enter a dormant state to wait for better times.

If the sphere were left in the dark, the recycling process would slow and soon come to a standstill. The larger organisms would die first. Bacteria would digest their remains, then they too would vanish and an equilibrium would be reached: a sterile, inorganic broth. But as long as light energizes the system and life is in charge, the medium, though itself not living, is a functional component. It is both a sew-

---

9.2 Cellular metabolism. This network is a biospheric elaboration of global geochemical cycles such as those shown in figure 4.1. *Adapted from Alberts et al.,* Molecular Biology of the Cell *(New York: Garland, 1983), p. 42.*

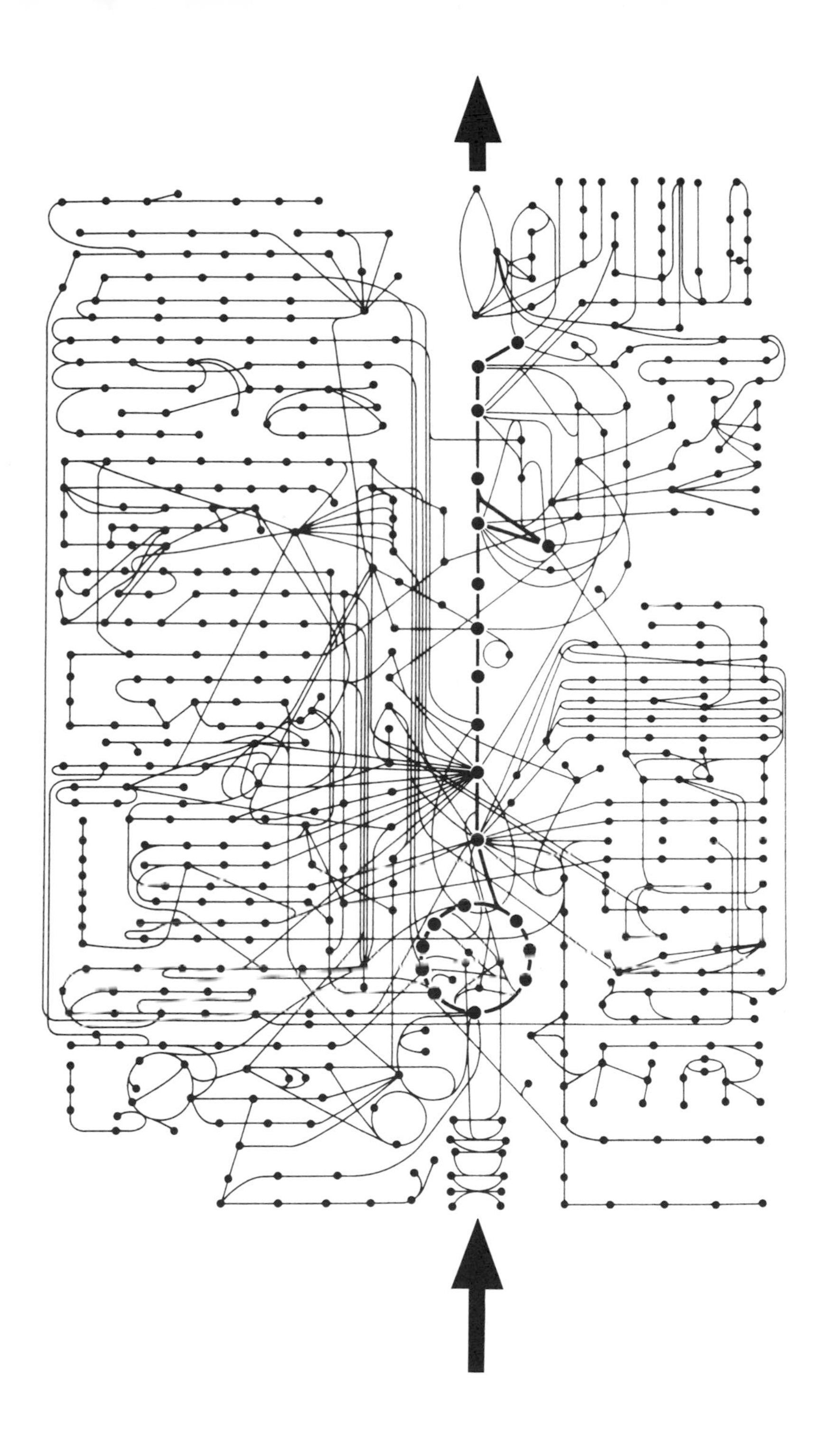

age depot and a continuous source of raw materials and nutrients, clear of toxins; a comfortable home, cleansed and organized.

Energy

Cleansing is work, and energy is required to carry it out. Unlike nutrients, energy cannot be recycled. It flows away dissipates, and is gone forever. Nevertheless, energy can be trapped and stored, and its flow can be conducted along intricate pathways and used for specific purposes. Water may be forced by solar radiation to evaporate and escape into the atmosphere. It may fall back as rain in the mountains and be collected in a reservoir lake. From there it may be conducted downslope to drive an electric generator. An electricity network may be activated, and an infinite variety of gadgets and tools set in operation. Ultimately, all that work, driven by the sun, is dissipated as heat.

So it is with the crystal sphere. All the living systems inside it are energized, and an elaborate network of energy pathways is woven throughout the living community, complementary to and entwined with the metabolic circuit.

Figure 9.1 shows in a basic way what is going on. The green algal cells collect some of the solar radiation that penetrates the sphere and utilize that energy to drive photosynthesis, the chemical reaction on which all further life depends. Light may be considered as a flux of particles, or photons. The green cells have minute funnels by which they can capture photons and direct them to reaction centers where their combined impetus drives the photosynthetic process. The two raw materials of photosynthesis, water (H_2O) and carbon dioxide (CO_2), are very stable molecules. But in the reaction centers these molecules are subjected to such an intense flux of photons that they are converted into a pair of very unstable substances: organics and oxygen (O_2). The energy of the photons is thus converted into chemical energy—a form that can be utilized by the algae themselves and by the shrimps and other organisms in the bowl.

When molecules of CH_2O and O_2 hit each other with force, H_2O and CO_2 are regenerated. The chemical energy (originally derived from the sun) is released as heat in the process, just as heat results when you burn paper or other organic matter. But life has more prudent ways to handle its energy than wasting it on fire. The transfer is gently conducted through a multitude of intermediate steps. At each step, some of the energy is utilized, some is passed on to be used at a larger stage, and a bit is lost as heat. It is only at the very end that water and carbon dioxide are finally regenerated. Ultimately, though, all the energy is dissipated as heat.

Comparing Ecosphere and Biosphere

Does the ecosphere on my windowsill help us to understand the biosphere at large? Intuitively, we feel that there is a close kinship between the two. The livestock in the ecosphere scavenges the nutrients and energy from the environment, and the nutrients are then used over and over again, while the energy is efficiently channelled through the system. Poisonous materials are diligently kept out of the way.

In the biosphere at large the same mechanisms appear to be in operation. Throughout the history of life, natural selection has favored the emergence of powerful biological "mining" mechanisms whereby nutrients are scavenged from rocks, water, and air. Indeed, we saw that living organisms are intimately involved in weathering and in the formation and maintenance of soils. Another point in case is iron, a very abundant metal on earth, but nevertheless terribly difficult to reach for most organisms. Iron rusts when exposed to oxygen, and rust is one of the most insoluble materials in existence. Many bacteria, however, excrete special organic molecules capable of plucking the iron out of the rust. Then they retrieve the molecules that have been successful in scavenging iron—and digest them. As a result of this mechanism, the biosphere is pervaded by a thin cloud of iron, secured by subtle biochemical mechanisms that evolved over the eons.

Once retrieved, nutrients are thriftily used and reused, especially when they are in short supply. This principle is perhaps best illustrated by tropical rain forests. These systems receive up to twenty feet of water per year, six times more than such a notoriously rainy place as the Netherlands. The nutrients are leached out of the soil by these large fluxes of water, leaving the soils impoverished. Yet there is no more luxurious blooming of life on earth, because the forests efficiently recycle the nutrients.

As for the removal of poisons, organisms used innumerable tricks to rid themselves of these unwanted materials. Poisons may be immobilized or pumped out of cells by specific macromolecules. They may be conducted to special tissues for storage or dumped into skeletons. Bacteria may trap heavy metals in their cell walls to prevent these toxic materials from entering the biochemical machinery inside. A large uranium spillage into a lake was once observed to be swiftly cleansed by a colony of diatoms. These tiny algae trapped the metal and, when they died, they settled on the floor of the lake, taking the uranium with them. This particular example shows that the detoxifying mechanisms may have repercussions far beyond the domain of the individual cell.

One must remember that a poison for one organism may be a much-wanted foodstuff for another. For instance, hydrogen sulfide (H_2S), the smelly gas emanating from stagnant waters and rotting eggs, is a common waste product of certain bacteria. For us it is extremely toxic, but there are many bacterial types to whom it is fresh air. They thrive on it and convert it into sulfate, an innocuous substance (at least for us). It appears that the solar energy trapped by plants and other photosynthesizing organisms is conducted with no less economy throughout living communities on the earth than inside the crystal sphere. So, the cleansing which we observed in the crystal sphere appears to pervade the entire biosphere. Can we the move our monastic experimental community into the mainstream of life? Can it help us to understand life at the planetary scale?

Differences

The differences are immense, despite the obvious similarities. When I first gazed into the sphere I was disappointed by the apparent insignificance of the organisms inside. But on second thought I changed my mind. Assume that the total biomass inside the sphere is one-thousandth the mass of the water and air in which it lives. By comparison, the amount of living material on earth is trivial. The mass of the atmosphere is about 2500 times that of the biota, whereas the hydrosphere (oceans, rivers, lakes, and ground water) is almost a million times bulkier. In the open oceans, the concentration of living matter is vanishingly small: it constitutes less than one-billionth of the total water mass.

The crystal sphere is a closed system for matter, the biosphere is more open. In the sphere, nutrient recycling and detoxification are entirely dependent on the organisms; on the earth at large these processes are integrated into a far larger cycle. Slowly revolving convection currents in the deep earth carry rocky materials from the surface into the interior and back again. A second, much faster system operates at the surface of the planet where the sun's radiation stirs up flimsy sheets of water and air. There is an elaborate system of routes for each substance, entwining and diverging, slowing down and speeding up, moving round. This is the cycling system of geochemical fluxes and reservoirs studied by Robert Garrels (fig. 4.1).

The geochemical cycles are at their most complex at the interface between rocks, water, and air. The fluxes of matter and energy are vastly accelerated, their course regulated by a self-propagating organization: life. This interface is the most active part of the biosphere. The metabolic network of a living cell shown in figure 9.2 is part of the geochemical cycles. Life is an elaboration of these streams of matter, itself a geochemical process. By contrast, life in the crystal sphere occurs in monastic seclusion, cut off from the main fluxes of matter.

Life in the sphere survives by rigorously organizing its secluded environment. On the earth at large, life must also cope with potentially overwhelming external pressures and changes. The thin soil layer on which terrestrial life depends is continually washed downslope; unpredictable accumulations of toxic materials may poison the environment, or slight changes in the cycles of the major elements may turn the earth into an uninhabitable desert. Life, suspended in the geochemical cycles, is at constant risk.

So, although the crystal sphere on my windowsill is strikingly similar to the real biosphere in some respects, we begin to discern fundamental differences. Instead of an isolated phenomenon, life on earth is an integrated component of a highly dynamic system of far greater size. One consequence of this coupling of geochemical and biological cycles is the amazing fact that the earth behaves as an enormous chemical battery, with the biosphere squeezed between its poles. Better than anything else, this condition demonstrates the important role that life has played in earth dynamics over the eons. Also, it adds to our doubts about the claim that closed ecosystems can be used as reliable guides to understanding life at the global scale.

Life itself has created this battery and depends upon it for its remarkable activity. One pole represents oxygen, which makes up 21 percent of our atmosphere. It is one of the most poisonous gases, reacting everywhere it penetrates: in soils, rocks, water, and air. Iron rusts, the organic materials of which living systems consist are converted into carbon dioxide and water, dead bodies disappear, and smelly gases are attacked and broken down. Oxygen is the major purifying agent of the outer earth. It cleanses the environment, gives us blue skies, and clears the air of smog and filth.

The other pole of the battery is in the ground, in the form of the huge reservoir of organic matter with which oxygen so readily reacts. At the interface between the realms of atmosphere and earth, a very thin rag of living organisms continually replenishes this energy supply, maintaining the battery's power and using the power for its own propagation. How, exactly, has our planet acquired this remarkable state?

Oxygen

The coupling of photosynthesis and respiration shown in Figure 9.1 forms the big flywheel of life. Carbon dioxide and water are converted into organic carbon and oxygen, and back again. The figure gives only a partial picture, however, since it suggests that the oxygen released by photosynthesis can immediately be used for respiration. In fact, respiration draws on the huge atmospheric oxygen reservoir, and without this reservoir the cycle could not operate on any significant scale. Oxygen accumulates in the atmosphere because a rapidly spinning biological flywheel is superimposed on a sluggish geochemical cycle involving sediments and rocks in the deep earth (fig. 9.3).

A tiny fraction of the organic carbon produced by photosynthesis escapes respiration. For every thousand molecules of organic carbon produced in the biological cycle, one molecule is buried in the sediments and transported into the deeper earth. This leak may seem trivial but once the organic carbon reaches the lithosphere, it remains there for a very long period, about 400 million years. As a result the mass of organic carbon stored underground is gigantic—10,000 times the total mass of all living things on earth.

Now, for every molecule of organic carbon withdrawn from the biological cycle into the lithosphere, one molecule of oxygen is added to the atmospheric reservoir. Of course, the story does not end there. If it did, the oxygen content of the atmosphere would still be increasing. Plate tectonics slowly brings rocks from the interior to the surface, where organisms catalyze a reaction between the uprising organic carbon and oxygen. Carbon dioxide and water are produced again, and in turn are utilized by the biological cycle. In other words, it is the combined effects of the biological cycle and the rock cycle that have made the global system. The biological cycle fuels the battery. The internal dynamism of our planet maintains the rock cycle, buries the organic matter, and sets the oxygen free. Without plate tectonics we would not be here.

This picture gains significance when we place it in the perspective of earth history. When life originated 3.5 billion years ago, there was little or no free oxygen in the atmosphere. Soon, photosynthetic organisms began to dump oxygen into the atmosphere, but it must have been a long time before sufficient organic matter was buried for the concentration of this gas to reach significant levels. In addition to the organic matter produced by life itself, the rock cycle brought up huge continuous fluxes of other mate-

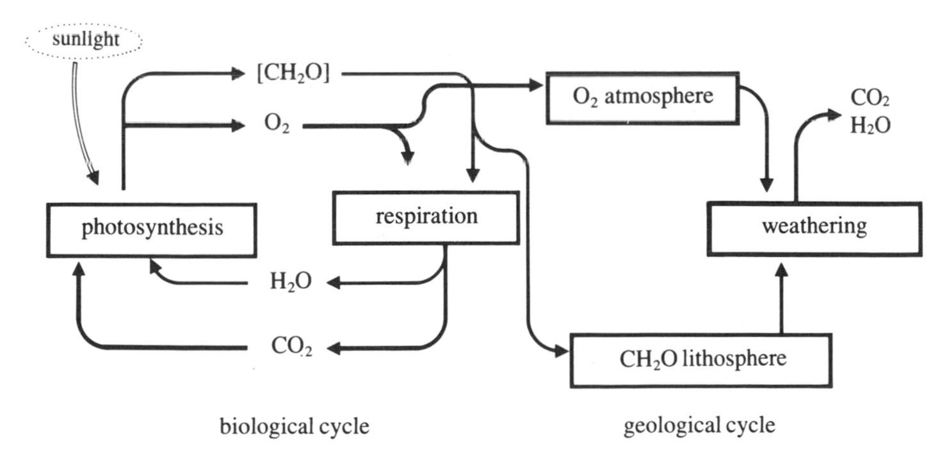

9.3 Coupling of the biological and geological cycles of organic carbon and oxygen: the origin of oxygen in the atmosphere.

rials—in particular ferrous iron and sulfides—that could react with oxygen. At first these substances scavenged away all oxygen from the air and in the process were transformed into huge masses of rust and gypsum. Then, about 2 billion years ago, these sinks were filled up and free oxygen began to accumulate.

This event must have been the greatest environmental disaster ever. Oxygen, a calamitous pollutant, made the atmosphere reactive to organic matter and poisonous to most life then in existence. Virtually all the existing biota were forced into sediments, stagnant waters, and other environments where this poisonous gas has no access. Some organisms, however, managed to survive the reactivity of oxygen, and others even "learned" to exploit it for energy. They transformed the peril of oxygen into a driving force of life on earth.

A Science of Life at the Global Scale?

Over the eons, a vast repertoire of mechanisms has evolved by which the retrieval and utilization of energy, the scavenging and recycling of nutrients, and the disposal of waste are secured by and for living organisms. The biosphere *appears* to be analogous to the crystal sphere on my windowsill—a world of its own, energized, organized, to some degree independent of the rest of the earth. But is it? Is the crystal sphere a reliable guide to the biosphere, sparking hopes for a genuinely global theory of life?

There are great differences. The fine tissue of life on earth is at constant risk, suspended in the network of global geochemical cycles. Over the eons, the internal dynamism of the earth and photosynthesis have joined forces to build a global geochemical battery through which the biological activity is vastly enhanced. This notion of gradual development and of close interaction between physical, chemical, and biological forces on earth brings home the idea that the biosphere is much more than any closed ecosystem. If at all, my crystal sphere only provides a poor analogy, however cosmic the complexities of its workings may be. However inspiring and instructive a deep look into its interior may be, any attempt to use it as a model for studying the biosphere at large must be viewed with suspicion.

The contents of my crystal sphere form a highly organized whole, acting as gracefully as a ballet dancer. This remarkable property of organization is proper to life. Wherever we look we see signs of it. But does the biosphere act in a similar way? Does it behave as a unified system, an organized whole, or is it no more than a loose amalgamate of living compartments, a fluid mosaic of organized parts? We don't know. But since we wish to understand the workings of our planet, we badly need an answer to this fundamental problem.

It now becomes apparent how misleading it is to apply the term *biosphere* to closed ecosystems. This delusive play of words is a good example of circular reasoning and wishful thinking. It merely suggests, but by no means proves, that the real biosphere behaves as a unified system. It obscures the issue by providing an a priori answer to the problem we wish to address. Only a top-down view of the earth itself can place the role of life into the correct perspective.

10

GAIA AND THE FRANKENSTEIN MONSTERS

JAMES LOVELOCK, inventor and independent scientist, challenges our notions of how the world works. His combined experimental station and home rests secluded in a pleasant valley in Cornwall. There, far removed from the regimentation of academic science, the intransigent experimentalist has developed a controversial view of the earth. In a hypothesis of startling simplicity and, some would say, in shocking disregard for established dogma, Lovelock argues that we should regard life not at the level of molecules, cells, organisms, or even ecosystems. Instead, the living organisms of the earth and their material environment together form a system that behaves like a single living entity. He calls this system Gaia, Mother Earth, after a goddess of ancient Greece.

Like Robert Garrels, whose reconstruction of the world is discussed in chapter 4, Lovelock wants to know how conditions on earth have remained stable enough that life could persist for more than 3.5 billion years. Unlike Garrels, however, Lovelock believes that if we view this planet as a chemical factory, we'll never understand how it works. Instead, he assumes that the operation of the Gaia system—all organisms and their environment together—is such that it automatically keeps our planet in a comfortable state, fit for the propagation of life.

Lovelock also tackles his work differently than do the geochemists. "The models they are using are totally inappropriate. They are collections of loose bits and pieces," he explained to me. "It's like assembling a Frankenstein monster. These modelers put together bits left over from dead bodies, and then they expect it to come alive. It won't."

In Lovelock's view, such models are based on a kind of mechanical logic that should have been thrown out of the window long ago. The real world operates in a more circular way; cause and effect are inextricably entangled. What he does in *his* models is try to create a simple, living, organized whole, instead of an automated monster.

Gaia in Action

In brief, this is the idea behind Lovelock's Gaia hypothesis: If you are a healthy person, your blood pressure, body temperature, and many other variables in your biological system are automatically maintained at close to optimum levels, regardless of changing conditions outside. People in biological and medical circles call the stable condition of healthy organisms *homeostasis*. Lovelock maintains that homeostasis is also an essential concept for understanding the earth. A network of Gaian regulatory mechanisms, operating at a global scale, keeps the earth in this remarkable state.

Recently, I explained the Gaia hypothesis to a professor of theology. "What is so special about this?" he asked. "To me, it seems very obvious." Indeed, to people with little training in the earth sciences but with a profound awareness of the surrounding world, Gaia may be a matter of course, but in the world of natural science the idea upsets all standing traditions. As I have maintained throughout this book, life represents a geological force and has deeply influenced the history of this planet. This view is increasingly acceptable to most scientists. But for Lovelock, Gaia is the master of all geological forces, and is in charge of the planet. Which professor of earth sciences would dare to make such a statement! In the face of much skepticism, Lovelock has drawn the notice of the worldwide scientific community, including a growing cadre of enthusiastic supporters who believe that Gaia contains the seeds of a new science of life on earth. Gradually, bits of data are accumulating that, although they don't prove Lovelock's hypothesis, tend to support its overall conclusions.

Lovelock is accused of teleology—the idea that planetary self regulation is purposeful or involves insight or planning by life. But, unperturbed by his critics, he follows his own instincts, deeply convinced that Gaia exists. He maintains that Gaia is a top-down systems view of the earth, the hard science of a physical chemist with an interest in control theory. Never some trendy New Age pseudo-science.

The Seeds of Gaia

Gaia, which deals so intimately with the earth, is a product of the science of space. Lovelock was involved with NASA when it was preparing for the Viking landing on Mars. The big issue was whether there was life on that planet, and if so, how it could be detected. Lovelock realized that the best way (if not the only reliable way) to detect life on an alien planet was to look at its atmosphere; there was no need even to visit the planet. All that is needed is a good infrared telescope.

One can imagine that this was not the most popular idea among the NASA officials who were in charge of the mission, but it turned out to be exactly right. On Mars and Venus there is neither life nor death, and their atmospheres are very close to chemical equilibrium, the condition you get in a closed container when all possible chemical change has taken place. A piece of wood in our atmosphere is not at equilibrium, but the ashes, water, and carbon dioxide that remain after it has gone up in flames are. The atmospheres of Mars and Venus are nearly all carbon dioxide, whereas the air we breathe is an anomalous mixture of oxygen, nitrogen, methane, with only a trace of carbon dioxide. The earth's atmosphere is incredibly reactive, at least on a geological time scale. An infrared telescope tells you the atmospheric composition of the planets. If you looked at the planets from outside the solar system with such an instrument, you would immediately notice the difference. In the atmospheres of Mars and Venus everything that could react has already reacted, but the earth, with its reactive atmosphere, is the only planet carrying life and death.

Being far, and persistently, removed from equilibrium is characteristic of life. Organisms use the atmosphere as a conveyor belt—a source of nutrients and a refuse sink—and that is what gives the air its highly unlikely composition. The other distinctive property of Gaia is its enormous resilience. Throughout 3.5 billion years the earth has been fit for life, even though the energy output of the sun has increased by more than 25 percent during that period. Gaia has survived terrible catastrophes. She has been bombarded by gigantic meteorites, has survived the buildup of oxygen (a terribly poisonous gas, as we saw in the previous chapter), and adapted to a dramatic decrease in the internal dynamism of the earth.

Yet never has the Gaian system departed from the narrow range of conditions required for the subsistence of life. The average temperature has always remained between 0° and 100°C; the concentrations of nutrients such as phosphorus, nitrogen, and sulfur have always been adequate; poisons have been sufficiently kept in check; the sea has never been more salty than organisms can withstand; oxygen has been maintained at tolerable concentrations; and water has been prevented from escaping the planet. Lovelock concludes that the climate and the chemical properties of the earth have been optimal at any time for the kind of life that inhabited the planet. For this to have happened by chance is as unlikely as it would be for a person to have survived unscathed a drive, blindfolded, through rush-hour traffic. Lovelock believes the concept of Gaia is the only plausible explanation.

Daisies

To clarify his understanding of Gaia, Lovelock invented Daisyworld, a planet ruled by mere flower power. Like all inventors, he doesn't care how it works, only that it works. A demonstration goes something like this.

Lovelock presses some buttons on his personal computer. Two coordinates appear. The horizontal axis gives the time, and the vertical axis gives the temperature on the planet. An imaginary planet spins round an imaginary sun. The star has increased its radiation over the imaginary eons, just as our own sun has done since its genesis. Near the origin of the coordinates a line appears, slowly extending towards the top right of the graph. It moves in a stepwise manner with small time increments. After each step, the computer calculates what the planetary surface temperature would be after the next increment and a new point is added at the end of the growing line. The surface temperature of the planet increases in a perfectly straight line: 0°C, 1°C, 2°C, 3°C, 4°C, 5°C, and then, all of a sudden, it makes a giant leap to 20°C. There it stays, for eons. Then it makes another leap upwards before continuing the original trend of a straight line. The result is in figure 10.1 (A).

Welcome to Daisyworld. At the beginning of the run, imaginary seeds of daisies are distributed over the planet. All daisies have identical growth characteristics: their seeds germinate at 5°C and the daisies die at 40°C. Their growth rate does not remain constant over that temperature trajectory: like all living systems in the real world, the daisies follow a bell-shaped curve. For the daisies, this moves upwards from 5° to 20°C, and then down again. All daisies are the same, with but a single difference: their shade. There are black daisies and white ones, and various shades in between. The darker daisies absorb more light, raising their own temperature and that of the planet. The white daisies reflect more light, cooling their environment. Before starting a run, Lovelock can indicate how many different types of daisies he wants, and also how dark or how light they are to be.

As soon as the daisies can begin to grow they influence the local temperature of their surroundings. The added effects of growth of all the daisies bring the planetary surface to a temperature where they grow best, 20°C. As in an individual mammal's body, the temperature is kept constant, in spite of external perturbations. Eventually, no life withstands the powers of the sun, the gradually increasing energy output of the star burns the daisies to death. This is Gaia is a nutshell: life and it surroundings create conditions on earth that correlate with this system's maximum proliferation, despite changing influences from outside.

Lovelock presses one of the buttons, and a second set of coordinates appears below the first one. Swiftly, the new graph is filled with a number of curved lines [fig. 10.1 (B)]. The left curve represents the black daisies, the right the white ones, and the others have intermediate shades, Lovelock explains. As soon as the critical temperature of 5°C is reached, the black daisies proliferate and entirely cover the planet. The surface is dark now, so that it retains the heat of the sun. Incidentally, you can also see that a few of the slightly lighter daisies come out—these reflect the solar radiation, thus lowering the temperature in their locality. Together, the dark and light daisies bring the surface temperature just to the optimum, and the relative abundances are neatly distributed so as to keep the surface close to that temperature.

On Daisyworld there is a tremendous bonus in diversity. The more shades of daisies you put in the model, the more precise the temperature regulation. When these thought experiments are undertaken, one suddenly realizes that this may explain why there is such unbridled diversity in nature, and how dangerous it is to wipe out species as we are doing today. As time elapses, and the intensity of the sun's radiation increases, the lighter daisies gain ground, gradually taking over from the dark ones. Finally, only the white dais-

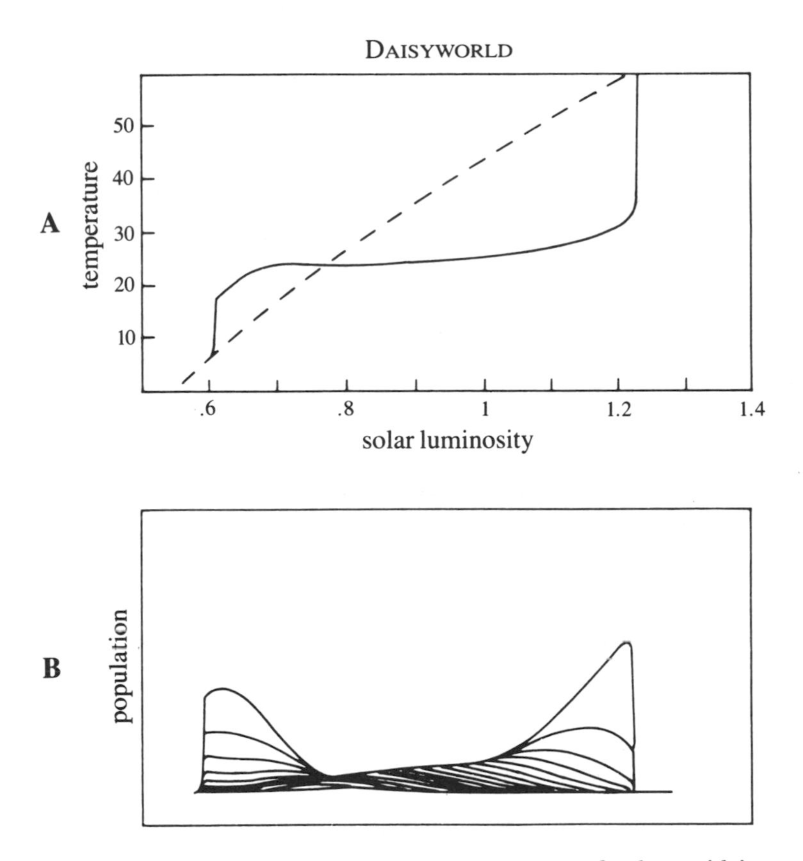

10.1 Daisyworld: (A) Change in surface temperature of a planet with increasing solar luminosity through astronomical time. Continuous line indicates the presence of daisies, the interrupted line their absence. (B) Change in the sizes of daisy populations during the computer simulation; dark daisies are prominent on the left and light ones on the right. *Adapted from James Lovelock,* The Ages of Gaia *(New York: Norton, 1988), fig. 3.3.*

ies remain. This goes on until the limit of the cooling capacity is reached; all daisies die, and again the planet behaves as a lifeless celestial body.

So, the system actively maintains a temperature which is optimal for its own propagation. It regulates the temperature over billions of years at the scale of an entire planet, without any foresight or planning.

Lovelock is the first to acknowledge that the concept of the earth as a superorganism did not originate with him. Nearly two centuries ago, James Hutton, one of the founders of geology, made a similar suggestion, and added that the appropriate method for studying our planet would be physiology—the science of the behavior of organisms. Vladimir Vernadsky, Arthur Redfield, G. Evelyn Hutchinson, and J. Z. Young also shared the general viewpoint. Lovelock has brought the old concept back in the language of today, so that we can no longer ignore it. After more than a century of stifling segregation between the sciences of life and earth, he reminds us of what we knew all along—that the distinction is not between physical, chemical, and biological forces, but between this strangely familiar planet and the rest of the solar system. He challenges the complacency of today's academia and makes us painfully aware of the need to reconstruct earth science.

Gaia versus the Frankenstein Monsters

Is geophysiology indeed the real science of our planet? Could Gaia chase the Frankenstein monsters of the geochemists? Let us lay the worlds of Garrels and Lovelock side by side and see how they compare.

Robert Garrels has given a global dimension to geochemistry by extrapolating from the established geological knowledge. He identified the major reservoirs and fluxes operating in the earth, and tied them up in a closed network of circulating pathways (figs. 4.1 and 4.3). Although he was deeply aware of the forces of life, his approach forced him to regard the big inanimate wheels of the system as the prime motivators of global dynamics, setting the stage for the deployment of the biosphere.

Lovelock starts from the opposite end. He explores the possibility that Gaia is in charge of the system. His alternative is Daisyworld, an imaginary, yet remarkable organized planet reflecting the immense power of self-sustenance for life. In contrast to the models of Garrels its purpose is not to simulate the actual world, but rather to demonstrate the basic principle of global homeostasis maintained by Gaia. For Garrels, the ultimate foundation and guidance of modeling was geological theory—the body of knowledge about the earth, gathered through centuries of painstaking research. For Lovelock, the point of departure is an idea. He believes that the geological data must be decoded and reinterpreted in light of his hypothesis before they can be used as elements of reliable models.

To probe fully the difference between the worlds of Garrels and Lovelock we may expose them to the rigorous rules of mainstream science. They both rely heavily on simplification and intuition, bring to the fore what is considered essential, and leave out the unimportant details. This is how it should be. Simplification and intuition are essential to all modeling and are indispensable for progress in science; but in themselves they are not enough. If we indulge in speculation for too long we find ourselves turning round in circles, and our effort is futile. Only when our ideas are tested in practice can we decide whether they are correct, and change them if necessary. On the other hand, experimentation and observation make no sense when left on their own. If these activities are not guided by theory they bog down into meaningless gathering of information, and there is no way to decide what that the information is worth. For progress in science, a tight coupling must be forged between theory and practice. The theory tells us what experiments and observations are worth carrying out, while the outcome of the testing furthers the development of the theory.

A tight coupling of theory and practice is only possible under very special circumstances. Any scientist can tell you that it may take many years before theory on the one hand and experimental and observational expertise on the other can be adapted until a close fit is reached. Out of the many ideas that we may develop only those that can be falsified are of value. This implies that the experimental techniques have to be developed until they allow the theoretical predications to be tested. The term *hypothesis* denotes a falsifiable scientific idea.

Garrels's work is an excellent example of this hypothetical way of progress. His fundamental hypothesis was that it makes sense to represent the earth as a chemical factory. He then designed a hypothetical earth factory of global chemical reservoirs linked up by a system of fluxes. Subsequently, the sulfur isotope curve demonstrated that there has been an antithetic breathing of the gypsum and pyrite reservoirs. Garrels realized that by itself this breathing made inadmissible demands on the reservoir of atmospheric oxygen. This finding allowed him to refine his hypothesis and make it testable. If it could be proved that the carbon and sulfur cycles are coupled so that the atmospheric oxygen remained constant, his representation of the earth was correct. When, at long last, the carbon isotope curve became available, he and Lerman could devise the computer model by which the coupling could be proved. The proof required a long search for the right questions and the right data to fit them. Even if it is a Frankenstein monster, Garrels's chemical factory provides us with a testable hypothesis.

And Gaia? When you look long enough at the crystal sphere on my windowsill with its enclosed community of shrimps, algae, and microbes you begin to wonder whether Lovelock is right, even if the biosphere at large is different in so many ways from this miniature bud. But is Gaia a hypothesis in the strict sense of the word? One must admit that the concept in its present form is rather fuzzy. For instance, Lovelock likes to compare Gaia with an organism, but in doing so he reduces the concept to an analogy, inappropriate for scientific testing.

What of the biological optimum condition for life that Gaia would maintain at a planetary scale? Contrary to the idealized Daisyworld situation, the concept of an optimum has little meaning in the real world. It can never be defined in terms of a single value for a single variable. In reality, there is a virtually infinite number of chemical, physical, and biological variables at stake. One might try to circumvent this problem by arguing that the optimum conditions for life as a whole are those that best increase its tendency to propagate. But this is only arm-waving. In the real world it is very hard, if not impossible, to establish experimentally what those conditions would be. For example, in a world without oxygen it would make little sense to say that this poisonous gas would bring the world closer to the optimum for life. And yet, the earth's atmosphere became loaded with oxygen.

Jim Kirchner, a geologist at the University of California at Berkeley, concludes on these and other, similar grounds that the Gaia concept cannot be falsified through experimental method. Because it offers a ready explanation for all phenomena imaginable, it explains nothing at all. Hence, Gaia is no hypothesis, in this strict sense of the word. Indeed, if the concept cannot yet be defined with sufficient precision, what is there to test? As much as I find the concept of Gaia attractive, for the practice of earth science its use is still limited.

Gurus

One need not decide whether Garrel's or Lovelock is right. Garrels himself had doubts about the potential of his approach to explain earth dynamics. He had demonstrated a coupling of the cycles of carbon and sulfur, but the mechanism underlying this remarkable correlation remained a mystery. Was his stylized world stable enough to support life for almost a stellar age? Despite its exemplary scientific methodology, his approach is founded on uncertain ground—a geological tradition that has all but ignored the forces of life for more than a century. The argument turns round in circles: models based on traditional geology will never bring the impact of life in perspective. Indeed, Garrels's models may well be Frankenstein monsters: they merely refer to an *aspect* of global dynamics, and appear not to reflect the behavior of the earth as a whole.

It is as if Garrels's models do not get off the ground, while those of Lovelock hover above the surface and cannot land. I cannot help being reminded of that other controversy that has occupied the minds of geologists for so long: the debate about stability and change. As we saw in chapter 3, until quite recently geological gurus were divided by apparently irreconcilable views of this problem. The French school interpreted the rock record in terms of long periods of stability interrupted by brief catastrophic events that fundamentally changed the earth. The English, in contrast, saw only gradual, directionless changes. They viewed the earth's surface as a fluid mosaic of separate environments: the pattern kept changing, but the elements remained the same.

The first glimpse of a resolution to this controversy only came about through the development of new technologies and concepts that could be applied to observations on a worldwide scale. Deep-sea drilling, and instruments that take accurate seismic readings of the ocean floor, provided a global perspective of sedimentary history. Marine paleontologists were able to provide the precise age determinations needed to distinguish between local and global events in the geological past. The respective arguments for stability and change became falsifiable. Finally, the geologists are able to see the patterns of change and to test, reconcile, and develop their concepts.

Today we are as ignorant about the role of life in global dynamics as the French and English geologists were about the nature of geological change. Are Garrels and Lovelock our gurus? More than anybody else they have probed the implications of a top-down view of the earth. It may be that their models, however revealing, are not good enough, and that making them bigger or adding them up will not help. A whiff of flower power will not bring the Frankenstein monsters to life.

Until the sciences of life and the earth are welded together, they will remain in a prescientific phase. They can only deal with parts of their subject matter, never the essence. Science is now concerned with plate tectonics, Garrels-type models, and life at the molecular, cellular, or ecosystem level. All these current fields ignore the vague but consistent assertion of the founders of geology—that life is a global system of self-organizing geochemical fluxes.

Until we can either identify an organizing principle that governs the behavior of the earth or refute the idea, we cannot tell whether information we gather about our planet is of any value in the clarification of this issue. Yet, in the absence of a sufficiently large body of information there is no hope that a falsifiable hypothesis will ever emerge. The prescientific level of the earth and life sciences forces us to grope, developing ideas and gathering information that we intuitively feel might be significant. One thing must be clear: it is not enough to study the earth as it behaves today. Only together will the present and the geological past provide sufficient keys for testing the models. We need to sharpen our perception of the biosphere's role in global dynamics today, and use that information to reconstruct from the rock record the evolution of life as a geological force.

11

EPILOGUE

GLOBAL CHANGE AND THE MELDING OF SCIENCES

FROM A BIOLOGICAL STANDPOINT the distance between humans and their closest relative, the chimp, is almost trivial, less maybe than between apples and pears. Yet a minute change in the genes marked a giant leap, a fundamental geological event, and a radical change of the face of the earth. I use the term *culture* to denote what distinguishes us from the natural world—not just a minor though significant departure in our genetic make-up, but also the ensemble of our organizations, utensils, institutions, arts, and ideas. I can understand but never sympathize with those biologists who still believe that our culture is only a flimsy varnish masking the naked apes we really are. Maybe that I live in an artificial landscape makes me particularly aware of the idiosyncrasy of the cultural world. I see culture everywhere and discern little, if any, untouched nature around me. Culture is our true habitat. We are cultural, not natural, beings.

When I look up from my work I see a desk covered with books, papers, pens, and an ashtray with a cigar stub. In my room are chairs, bookshelves, paintings, an old chest, a radiator, and a few potted plants. I hear music: sound organized by cultural forces, and designed to organize my emotions. Through the windows I can see the street, some gardens, and the houses of my neighbors. My environment is artificial as far as the senses can perceive. The trees on the street have been planted, as have all the woods in this country. Even the air that I breathe has been interfered with by humans. How far must I travel to find pure, unbridled nature, free of human influence? If you take into account our impact on climate, the oceans, and the atmosphere, nothing is unaffected. Even deep in the crust we tap reservoirs of water, fossil fuels, and raw materials on a huge scale. In space, you would have to travel far away to escape the cacophony of radiowaves emitted by civilization. The environmental impact of humanity is ubiquitous. We are immersed in culture, and there is no escape.

It was the French philosophers Édouard Le Roy (1870–1954) and Pierre Teilhard de Chardin (1881–1955) who first coined the term *noosphere* (from the Greek *νοοσ* or "mind") to denote the idiosyncracy and the planetary dimension of culture. Vladimir Vernadsky (1863–1945), the Russian earth scientist who in the first decades of this century developed a visionary and embracing concept of global dynamics, was in close contact with Teilhard. He regarded the noosphere as the envelope of the mind that was to supersede the biosphere, but in addition he emphasized the basic unity of culture and the natural world. In Vernadsky's view, the noosphere emerged from the biosphere and is an elaboration of it. Or (to use Lovelock's terminology), *Gaia* only refers to the physical, chemical, and biological underpinnings of the present earth's dynamics—to a pristine and bygone phase of our planet's evolution. Then *νοοσ* takes over, it seems, but in operation it is rather less august than the lofty Greek word suggests.

> From the paleontological and biological point of view, adjusting our gaze to that time scale, we see on Earth a gradual growth of the animal and plant life developing into ever newer, richer, higher and more perfect forms, until suddenly the development ends because with stupendous speed this monkey breed rises to divine power and becomes master of the Earth. Master of the Earth indeed, for now possession could be taken of the whole world.*

Less than forty years after these words were written by Anton Pannekoek, a Dutch philosopher of science, our view of the earth has dramatically changed. The notion that evolution is synonymous with progress is discredited, and we feel ourselves victims rather than masters of this process. Impending catastrophe?

Figure 11.1 sums up the development of humanity as a very slow and gradual progression, at first embedded in nature. Then from the Renaissance onwards there is a sudden, dramatic upsurge. The figure shows (from left to right) world population, energy consumption per capita, number of scientific journals, and mobility all increasing exponentially or even worse. And these are only four out of many

*Anton Pannekoek, *Anthropogenesis: A Study of the Origin of Man* (Amsterdam: North Holland Publishing Company, 1953), 90.

variables that appear to explode. We are hitting a wall, it seems, massive and ominous, barring any further expansion and positive action. This figure dramatically brings to the fore the anxious premonitions of our times. Some shut their eyes to the times yet to come, while others seek consolation in a weird brand of Gaia theory cum religion.

The end of the world has been foretold throughout written history. Forebodings of disaster fall on particularly fertile ground when societies are going through a critical developmental phase. Blind fear may spread as an epidemic to the farthest corners of the population. When scientists prophesy doom, their authority gives special weight to the feeling that disaster is imminent. Calamity is news, especially when it affects the world at large, and these days news spreads rapidly. Moreover, we must bear in mind that dramatic curves such as those in figure 11.1 highly stimulate the flow of money to the institutions responsible for these representations. Large sectors of scientific and technological endeavor thrive on our collective fears.

This creates a dangerous situation. In my native Holland, the public is increasingly willing to make great sacrifices, and this forces the politicians to take action. Proposals are hastily prepared to halt environmental degradation. Cars must be banned, industries liquidated, factories closed, and farming curbed! If, afterwards, such measures proved unnecessary an enormous backlash would ensue. People would no longer be prepared to make any sacrifices, even those really needed.

One should be grateful for the wave of environmental concern that is sweeping through our societies. It may be the beginning of a new relationship between humanity and the environment. But at present this concern is an untamed force. Not only do we have to understand the workings of our planet, but we must evaluate the consequences of our actions and reach a sensible environmental policy. To do this we must expose the sources of our collective feelings and beliefs.

A Sparkle of Hope

One offshoot of culture behaves in an unusual fashion. Time upon time, science liberates itself from the stranglehold of petty self-indulgence and private interest. At its

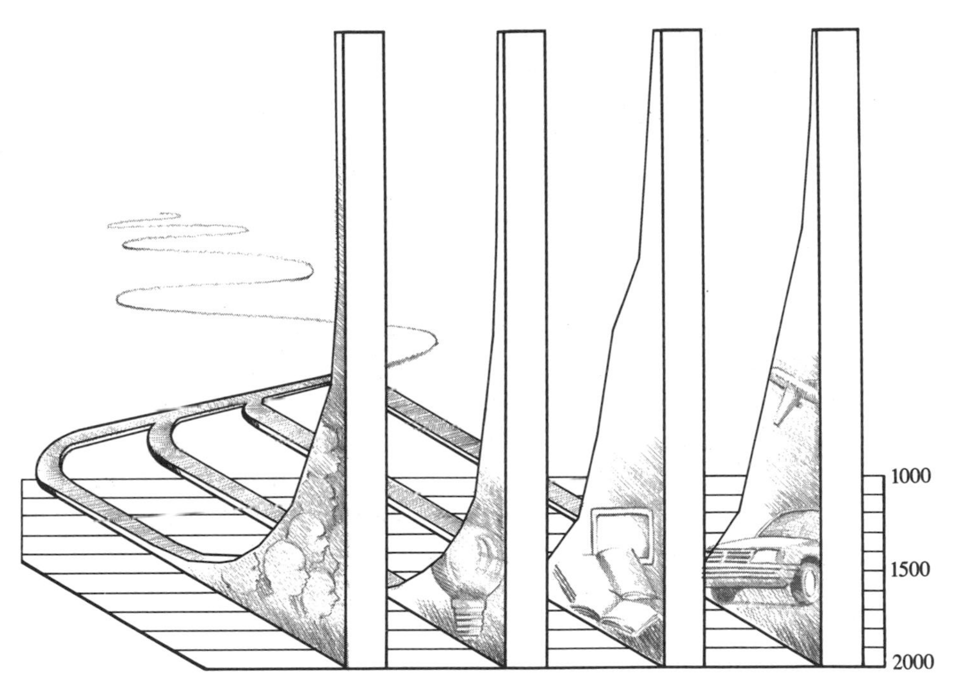

11.1 Explosive growth of cultural forces over the last thousand years. From left to right, increase in population density, energy consumption per capita, numbers of scientific journals, and speed of transportation. *Redrawn from* The Gaia Atlas of Planet Management, *by Norman Myers. Produced by Gaia Books, London, and published by Doubleday, New York.*

best, science is the most democratic and progressive of all human endeavors, breaking through all frontiers and recognizing no authority other than reality. Its powers are immense and, ideally, its ways cunning, patient, and careful.

Science offers no more than a sparkle of hope, because even in science we hit upon our tragic ignorance. If the earth and life sciences are immature, how shall we cope with culture? For the moment, *noosphere* is only a descriptive term with little scientific meaning. Gurus abound in the social sciences, but we lack a comprehensive theory that explains the integration of culture and natural science as a global phenomenon. And so, although we are aware of the environmental problem, we cannot grasp its essence.

The prescientific level at which the sciences of global dynamics are now stuck forces us to develop ideas and gather information intuitively. This is the policy of the giant effort that I briefly introduced at the beginning of this book: The International Geosphere-Biosphere Programme: A Study of Global Change. The objective of the program is "to describe and understand the interactive physical, chemical, and biological processes that regulate the total Earth system, the unique environment that it provides for life, the changes that are occurring in this system, and the manner in which they are influenced by human activities," according to one of its many colorful brochures.* One project, just started, illustrates how the environmental problem spurs integration of sciences that have long been apart.

Remember *Emiliania*, the wonderful cell with an elegant coat of coccoliths (from chap. 6)? That organism turned out

*International Geosphere-Biosphere Programme (IGBP), *A Study of Global Change* (Stockholm: The Royal Swedish Academy of Sciences, n.d.), 2.

to be a phenomenon of considerable geological and ecological importance. It is the major limestone-producing organism on earth, one of the most abundant species in the oceans, a key component of the marine food web, and it probably induces white clouds over the oceans. The name *Emiliania* pops up time after time in scientific meetings and in the literature, and probably no laboratory in the world has more experience with this species than ours.

And yet, how little we know. Why do the cells surround themselves with coccoliths? Some believe that these particles help the cells settle from the uppermost layers of the ocean towards deeper and more nutrient-rich waters. Others suggest that, by reflecting the light, they extend the length of the light's path through the surface water and enhance the absorption of energy. This would bring the water temperature closer to *Emiliania*'s growth optimum. How do these cells behave with respect to the huge water masses that subside into the depths at one place and well up to the surface at another? We don't know the exact extent of carbon-dioxide buffering by coccoliths and other floating lime-producers in the ocean, nor the importance of this mechanism as a component of the multiple systems that modulate carbon dioxide in the atmosphere. We are insufficiently informed as to the distribution and life cycles of coccolith-producing cells and the correlation of these attributes to climates and ocean currents, past and present. Does *Emiliania* form one worldwide interbreeding family, or are there separate strains in the different water masses and oceans?

In working with climatologists now, and with the Gaia hypothesis in mind, James Lovelock is exploring the significance of *Emiliania* as a producer of glaringly white clouds over the oceans, clouds capable of reflecting solar radiation.

Could these tiny organisms be the white daisies of Gaia? This brings to my mind the Cretaceous period, some 100 million years ago, the time when coccoliths accumulated in the oceans in gigantic amounts. It has been suggested that the concentration of carbon dioxide in the atmosphere was at least ten times its present values. The global climate was much warmer than now, and forests thrived on Antarctica. Maybe *Emiliania*'s ancestors enveloped the earth in a shield of white, cooling clouds, thereby compensating for the enhanced greenhouse effect of the high carbon dioxide levels.

Not long ago, I had the idea of launching a large research collaboration encompassing biochemical, physiological, ecological, and geological aspects of the *Emiliania* phenomenon. This organism would become a model system for both oceanography and geology, a key to understanding the impact of plankton on the household of the oceans, the atmosphere, the clouds, and the world's sediments. I have been surprised by the enthusiastic response of my colleagues, who proposed incorporating this project into global oceanographic programs. A large international team has now been formed that will participate in research cruises, deploying sediment traps in the deep sea and taking core samples from the ocean floor. The dynamics of *Emiliania*'s blooms will be investigated by integrating satellite imagery with detailed studies on ships and in the laboratory. Cells of the alga will be collected from all over the oceans. We will investigate the downward fluxes of coccoliths and organic materials as well as the emanations of the cloud-producing gas, dimethyl sulfide. Antibodies will be used to follow in detail the complex macromolecular signals emitted by the blooms towards the ocean floor. The sediment samples will reveal the history of the system during the last great climate cycles of the ice age. And last but not least, the dynamics of the entire system will be simulated in computer models.

By focusing on a single key organism, *Emiliania,* we intend to reach a new level of integration between the earth and life sciences. This initiative is a modest seed for more innovative studies. I hope that more and more scientists will discover how their knowledge can be made instrumental for understanding the peculiar planet on which we live. However, one must bear in mind that this kind of effort will only highlight aspects of earth dynamics, and by itself will fail to reveal the overall patterns of change. In particular, this work will be all in vain if in the end the sciences of nature and culture remain divorced.

The Earth of the Future

At the close of this book I think back to Nieuwkoop and its 800 years of environmental management. The area had been converted from an inhospitable wilderness into an artifact, an environment adapted by human shrewdness to human demands. The inhabitants had to cope with frightening environmental problems. Yet these were solved, time upon time, albeit often at a great cost. Now the problems are manifold and of global extent. Humankind may be overwhelmed, and the natural forces may at long last prevail again. But I feel that if we manage to survive, as an inevitable consequence the entire earth will be converted into a cultural domain—a kind of park, a giant laboratory, or a great work of art.

The earth will be tamed, but will it be tame? Think of New York, Paris, Moscow, or Cairo. Consider the mountains of Switzerland, the natural parks of the United States, the Alhambra in Spain, the Great Wall of China; think of music, architecture, art. You will agree that the noosphere can be majestic and compelling. No doubt the taming of the earth will create new problems. But our offspring will never have reason to be bored.

SUGGESTIONS FOR FURTHER READING

THOSE WHO WISH to be informed about the latest scientific developments in the various fields discussed in this book should scan general journals such as *Nature, Science, Scientific American, American Scientist,* and *New Scientist,* and pick up what is of interest. Reference lists and book reviews provide further guidance to the scientific literature.

Excellent textbooks are available with introductory systematic overviews of geology, microbiology, ecology, and other relevant fields. Tjeered H. van Andel's *New Views on an Old Planet: Continental Drift and the History of the Earth* (Cambridge, Eng.: Cambridge University Press, 1985) gives a fine introduction into current geological thought. The geologic history of the earth, with special emphasis on paleontology, is beautifully described by Steven M. Stanley in *Earth and Life Through Time* (New York: Freeman, 1986). A classic overview of earth history from a chemist's point of view is *The Chemical Evolution of the Atmosphere and Oceans* by Heinrich D. Holland (Princeton, N.J.: Princeton University Press, 1984). To become acquainted with microbial life, I much enjoyed reading *Biology of Microorganisms,* 3d edition, by Thomas D. Brock (Englewood Cliffs, N.J.: Prentice Hall, 1979), *Garden of Microbial Delights* by Dorion Sagan and Lynn Margulis (Boston: Harcourt Brace Jovanovich, 1988), and *Microbial Geochemistry,* edited by W. E. Krumbein (Oxford, Eng.: Blackwell, 1983). W. T. Keeton and J. C. Gould give an admirable overview of the life sciences in their *Biological Science,* 4th edition (New York: Norton, 1986). These are only a few of the multitude of texts available. For the latest and most relevant books on the market I recommend the reader to consult the experts in systematic teaching—university professors.

A beautiful account of Soviet thought and research on the role of life in global dynamics is A. V. Lapo's *Traces of Bygone Biospheres* (Oracle, Ariz.: Synergetic Press, 1986). This little book is also the best introduction I know into the work of the founder of biogeochemistry, Vladimir Vernadsky.

For those who have a chance to visit the Netherlands I recommend the excellent book in English by Audrey M. Lambert, *The Making of the Dutch Landscape: An Historical Geography of the Netherlands.* (London: Seminar Press, 1971).

I particularly enjoyed reading James Lovelock's books on his divine invention, *Gaia: A New Look at Life on Earth* (New York: Oxford University Press, 1979) and *The Ages of Gaia* (New York: Norton, 1988). Related in outlook, and emphasizing microbial life, is Lynn Margulis's and Dorion

Sagan's *Microcosmos: Four Billion Years of Microbial Evolution* (New York: Summit Books, 1986).

An outstanding reference on man-made environmental changes and the effects of these changes on human beings is provided by a series of workshop proceedings issued by the Scientific Committee on Problems of the Environment (SCOPE), and published by Wiley. In addition, the September 1988 issue of *Scientific American* deals with managing the global environment.

Probably the best advice I can give to those who wish to familiarize themselves with the problems discussed in this book is to go out and study their own surroundings, as I did in Nieuwkoop and Vlieland, or to join field excursions organized by experts.

INDEX

Page numbers in ***boldface italics*** refer to illustrations.

Abram, David, 184
acid rain, 146
acoustic beacons, 76–77
algae, 30
  in coral reefs, 156–58, 163
  in desktop biosphere, 183–84, 186–90, 192
  diatoms, 127, 133, 167, 174, 196–97
  importance of, 88
  Melobesia, 152, 153
  in microbial mats, 118–19, 126
  see also Emiliana
amoebas, 174
animals:
  as eukaryotes, 174
  newness of, 175
Antarctica, 226
antibodies, 150, 167
Appalachians, 62
arsenic, 189
asthenosphere, 49
Atlantic Ocean, 61
  tectonic spreading of, 49–50
atmosphere:
  chemical composition of, 207
  global cycling and, 94–96, ***95, 99,*** 100–107, ***104,*** 113, 196
  of Mars and Venus, 207
  mass of, 197
atolls, 158–62, ***161***

bacteria, 163, 166, 169–73
  as biosphere cleanser, 173, 186–87, 188–90

bacteria (*continued*)
  in desktop biosphere, 186–90, 192
  DNA transfers by, 173
  as earth's first inhabitants, 171, 175
  Escherichia coli, 146
  eukaryotes dependent on, 174
  iron and, 196
  in limestone, 148
  microbial mats of, 117–21
  in mud mounds, 175–79
  multiplication rates of, 190
  as one organism, 173
  in plant decomposition, 29
  structure of, 170
  sulphide derived from, 93
baggerbeugels, 37, ***38***
Bahamas, 132
basalt, 49, 52, 77, 92
beacons, acoustic, 76–77
Belgium, 176
Berner, Bob, 105
biological cycle, ***95,*** 96
biology:
  geology integrated with, 21–22
  small-scale view of, 68
  see also life force
biomass, 197–98
biospheres, 183–203
  bacteria as cleanser for, 173
  Biosphere II, 185
  in colonization of Mars, 185
  definition of, 185
  desktop, 183–200, 202–3, 215
  geosphere and, 146
  global cycling and, 106–7
  iron cloud of, 196
  noosphere and, 220–21
  slime as glue for, 167
Biosphere II, 185
Bluck, Brian, 58, 62, 63, ***67***
blue-green algae (cyanobacteria), 118-19, 126
body fluids, 154
brachiopods, 60, 61

cadmium, 189
calcite, 137–46, ***138***
  Emiliania's formation of, 137–46, ***138***
  polysaccharides and, 139–40, ***142–43,*** 155
calcium, 188
calcium carbonate, 93, 150, 174
  see also calcite; coral reefs; limestone
calcium chloride, 136–37
calcium phosphate, 154, 155
Cambrian period, 100, 105, 163
canals, 23, 24, 36, 37, 38
Cape Hatteras, N.C., 74, ***75,*** 76
carbon, 130
  desktop biosphere cycling of, 188–89, ***189***
  organic isotope of, 93, ***99,*** 100, 102, 103–107, ***104,*** 128, 132, 188–89, ***189***
  in radiometric dating, 79
  sulphur cycle coupled with, 102–6, 214, 216
carbonate platforms, 158–62, 163
carbon dioxide, 119, 132, 155
  in atmosphere, 207, 226
  in desktop biosphere, 187, ***189,*** 194–95
  in Garrels's and Perry's geochemical factory theory, 91, 93, ***99,*** 101
  global cycling and, 94–96, ***95,*** 173
  as greenhouse gas, 145
  in photosynthesis, 96, 101, 155, 194–95, 200–201
  in plant decomposition, 29
cattle, 37, 39
CELSS (Controlled Ecological Life Supports Systems) program, 185
cellular metabolism, 191–94, ***192,*** 198
chalk mud, 149, 150, 177–78
  chemical composition of, 151
  inorganic precipitation of, 153–54
  photosynthesis and, 155–56
  precipitation inhibitors for, 154–56
  sea grass and, 152–53
cheese, 37
chitons, 169
chlorophyll, 118, 119
chloroplasts, 170
climate, 23
  in Cretaceous period, 226
  Emiliania and, 144–46, 225–26
  ocean levels and, 85–86
  water cycle and, 129–30
cloud formation, 130, 146, 225
coal, 93, 132, 149

coccoliths, ***80,*** 88, 147, 164, 174, 225, 226
  definition and chemical composition of, 79, 137, ***138,*** 149
  formation of, ***142–43,*** 144, ***144***
  ocean floors covered by, 141
  polysaccharides and, 139, ***142–43,*** 155
  vesicles for, 137, 139, ***142–43,*** 144, 170
cockles, 127, 167
conglomerate, 55, 58
continental margin, passive, 74–75
continental rise, 76
continents:
  tectonic creation of, 52, 54
  see also specific continents
Controlled Ecological Life Support Systems (CELSS) program, 185
copepods, 141
copper, 134, 188
Coquand, H., 70, 84
coral reefs, 132, 141, 156–58, 163, 177
Cretaceous period, 158, 226
crystals, 136–37
culture, 219–22
  as distinguishing human characteristic, 219–20
  evolution of, 221–22, ***223***
  geological impact of, 31–39, 227
  science and, 224
Cuvier, Georges, 70
cyanobacteria (blue-green algae), 118–19, 126

Daisyworld, 208–12, ***211,*** 213, 215
dating techniques, 78–81, ***80,*** 217
de Boer, Poppe, 127
deep-sea drilling, 76–78, 217
Deep Sea Drilling Project (DSDP), 76–78, 83
desktop biospheres, 183–200, 215
  biomass of, 197–98
  chemical elements in, 188
  energy network in, 194–95
  metabolic networks of, 192
  microbial variety and functions in, 188–90
  natural selection in, 190
  photosynthesis in, 186–87, ***189,*** 190, 194–95, 197
  planetary biosphere compared with, 195–200, 202–3
  Russian doll metaphor for, 190–91
  shrimp and algae symbiosis in, 186–87
diatoms, 127, 133, 167, 174, 196–97
digestive tract, 167–68
dikes, 23, 24, 26, 36, 37
dimethyl sulfide, 146, 226
diversity of species, 211
DNA, 170, 173, 174
doom, prophesies of, 222
Dordogne Valley, 69–70, 84
drilling, deep-sea, 76–78, 217
DSPD (Deep Sea Drilling Project), 76–78, 83
dunes, *see* sand dunes
dust bowl, 131

earthquakes, 48, 50, 51, 53
EcoSphere Associates, 184
Emiliania, 137–46, ***138,*** 147, 155, 164
  blooms of, 140–41, ***145***
  calcite formed by, 137–46, 174
  climate affected by, 144–46, 225–26
  cloud formation and, 146, 225
  importance of, 140, 225
  mineral coat of, 141, 144
  research on, 224–27
  structure of, 170
environmental concern, 222–23
enzymes, 191
erosion, 38, 114
  life forms and, 129, 131–32, 134
  nutrients and, 85
Escherichia coli, 146
eukaryotic cells, 94, 170, 173–75
Europe, 61–62, 63
evolution:
  cultural, 221–22, ***223***
  ocean levels and, 86
  as progress, 221
  symbiosis and, 169–73
Evolution of Sedimentary Rocks (Garrels and Mackenzie), 89

fertilizers, 40
Florida Bay, 151–53, 154, 155, 156, 158, 159, 167
Florida Keys, 156

foraminifera, 79, ***80,*** 88, 153, 174
forests:
in Cretaceous period, 226
depletion of, 114
rain, 129–30
fossils, 94
off Cape Hatteras, 77
coral, 176
in France, 70, 84
importance of, 67
in radiometric dating, 78–81, ***80***
in Scotland, 60–61
sedimentary, 149–50
total accumulated organic material of, 150
France:
catastrophic geological school of, 70–72, 81–82, 84, 216–17
Dordogne Valley of, 69–70, 84
fungi, 130, 131, 172
in plant decomposition, 29

Gaia hypothesis, 204–18, 221
Daisyworld computer model of, 208–12, ***211,*** 213, 215
definition of, 204–6, 210
nonfalsifiability of, 216
homeostasis in, 206, 213
origins of, 207–8, 212
orthodox science and, 206
as religion, 222
resilience in, 207
Garrels, Robert, 89–107, 113, 198, 205
biological cycling ignored by, 96, 213
carbon and sulphur cycles coupled by, 102–6
death of, 91
dream inspiration of, 105–6
geochemical factory theory of Perry and, 91–93, ***92,*** 97–102, ***99,*** 214–15
geochemistry established by, 89–90
Lovelock compared with, 212–17
geochemisry, 21, 89–107
Garrels and, 89–93, 96–107
Garrels's and Perry's chemical factory theory of, 91–93, ***92,*** 97–102, ***99,*** 214–15
global cycling and, 94–107, ***95, 99, 104***
geology, 45–107
biology integrated with, 21–22
catastrophic school of, 70–72, 81–82, 84, 216–17
change in, 85–88
emergent disciplines of, 21
global-scale view of, 21, 47–48, 65, 89, 111–13
gradualist school of, 71–72, 82, 84, 216–17
of Java, 51–54, ***53,*** 63, 64
life force ignored by, 21–22, 68, 88, 216–18
of Scotland, 54–64, ***55, 56, 64***
see also plate tectonics
geophysics, 21
global cycling, 94–107, ***95, 99, 104,*** 113, 173, 198
breathing analogy for, 98–102, ***99***
carbon and sulphur coupled in, 102–6
rigid vs. flexible models of, 97–98
time scales for, 94–96, ***95***
Glomar Challenger, 76
gold, 134
granite, 52, 58, 63, 92
grasses, 128, 130
sea, 150–53
of Vlieland's sand dunes, 121–26, ***125***
Great Glen Fault, 62
greenhouse gases, 145
Greenland, 62
gypsum, *see* sulphur, sulphate isotope of

heliozoa, 174
Highlands, 62, 64, ***67***
Holocene period, 26, 28, 30
homeostasis, 206, 213
Hutchinson, G. Evelyn, 212
Hutton, James, 113, 212
hydrogen, 188
hydrogen sulphide, 173, 197
hydrosphere, 197–98
hypotheses, falsifiability of, 214, 216

Iapetus Ocean, 61–62, 63, 64, 66
ice ages, 26, 28, 114, 227
Indian Ocean, 51
International Geosphere-Biosphere Programme (IGPP), 41, 224
Ireland, 62

iron, 134, 196, 199, 202
iron sulfide, 93

Jan de Bruyn, Gerrit, 121
Java, 51–54, ***53,*** 63, 64, ***67***

Kirchner, Jim, 215–16

lagoons, 22–23
lead, 189
Leg 93, 77–78, 83
Lehninger, Albert, 154
Lerman, Abe, 103–6, 215
Le Roy, Édouard, 220
lichens, 128, 130
life force:
  geology and, 21–22, 68, 88, 216–18
  global cycling and, 107, 113, 199–200
  growing evidence of, 21–22
  in Iapetus Ocean, 66–67
  limestone evolution and, 147–65
  Netherlands landscape shaped by, 31–42
  rock cycling and, 129–35
  in sand dune cycles, 128–29
  sediments and, 132–34, ***133***
  soil erosion and transport and, 131–32, 134
  water cycle and, 130
limestone, 147–65
  in Belgium, 176–79
  biochemical approach to, 148–50
  chemical composition of, 93, 137, 149
  evolution of, 162–65
  in France, 69, 70
  life force and, 147
  organic carbon isotope derived from, ***99,*** 100, 101, 102, 103–107, ***104,*** 132
  precipitation inhibitors for, 154–56, 167
  in rock cycle, 162–63
  in Scotland, 58, 61
  sea grass and, 150–53
  see also calcite; calcium carbonate; coral reefs
limpets, 169
lithification, 129
lithosphere, 48–50, 90, 91
  definition of, 48–49
  global cycling and, 94–96, ***95,*** 100–107, ***104***
Lock Ness, 62
Lovelock, James, 204–18, 221
  Daisyworld computer model of, 208–12, ***211,*** 213, 215
  Emiliania research of, 225–26
  Gaia hypothesis of, 204–8
  Garrels compared with, 212–17
Lower Cretaceous period, 77
Lower Pleistocene period, 77
Lyell, Charles, 71

Mackenzie, Fred, 89
McKerrow, Stuart, 62, 63
macromolecules, 196
  antibodies, 150, 167
  calcium carbonate and, 150
  organic fossil, 150
  slime, 131, 148, 150
magma, 52, 58
magnesium, 154–55, 188
magnesium carbonate, 93
manganese, 134
mantle, 49–50, 65
Margulis, Lynn:
  on bacteria and evolution, 170
  Gaia hypothesis and, 206
  microbial mat studies of, 172
  on symbiosis, 170–71
marine snow, 175
marram grass, 123–26, ***125,*** 132
Mars:
  atmosphere of, 207
  proposed colonization of, 185
  rocky surface of, 167
  Viking mission to, 207
Mauritania, 62
Meije stream, 23–24, ***25,*** 31, ***34,*** 36, 39
Melobesia, 152, 153
mercury, 189
metamorphism, 96
methane, 173, 207
Meuse river, ***27,*** 28, 114
microbial mats, 117–21, 163, 178
  bacteria in, 118–21, 163
  definition and chemical composition of, 118–21
  density of, 119–20, 126
  diversity in, 119–20, 172
  moisture maintained by, 117–18, 120, 129

microbial mats (*continued*)
slime in, 129, 167, 175, 177
Microcoleus, 172
Mid-Atlantic ridge, 49
Midland Valley, 62, 63
mitochondrion, 170
mollusks, 169
molybdenum, 188
Mondriaan, Piet, 91
Monty, Claude, 177–79
mosses, 128, 130
sphagnum, 28, 30–31
mountains:
rocks weathered from, 57–58, ***59,*** 129
see also rock cycling
mucus, 167
mud chalks, *see* chalk mud
mud mounds, 175–79
mussels, 167

National Aeronautics and Space Administration (NASA), 185, 207
natural selection, 190, 195
Netherlands, 22–39, 42
culture's geological impact in, 31–39, 227
environmental destruction, 38–39, 40–41, 222
geological history of, 26–28
Golden Age of, 26, 38
land reclamation in, 34–37, 40–41
landscape of, 22–26, 39–40
peat formation in, 28–31
Nieuwkoop, 22–26, ***25, 27,*** 31, 34, 39, 40, 227
nitrate, 40, 162, 169
nitrogen, 119, 128, 130, 173, 188, 207, 208
noosphere, 220–21, 224, 228
North America, 49, 61–62, 63
Norway, 62
nutrients, 119, 162, 208
in cellular metabolism, 191
poison from high concentrations of, 189
recycling of, 186–87, 196
in rock cycling, 127–29, 130, 135, 195–96

oceans:
ancient, 61–62
changing levels of, 28, 78, 82, ***83,*** 84–85
climate and levels of, 85–86
crust under, 52
global cycling and, 91, 93, 94–96, ***95,*** 101–107, ***104,*** 113
marine snow of, 175
nutrient levels of, 85–86
oxygen levels of, 85–86
sediment in, 52, 53
see also subduction zones; *specific oceans*
oil, 72–73, 93, 132, 149
Ordovician period, 54, 58, 60, 61, 64, 66
oxygen, 196, 200–202, 207
cyanobacteria's atmospheric introduction of, 119, 200–202, ***201***
in desktop biosphere, 187, 189
in Garrels's and Perry's geochemical factory theory, 91, 93, ***99,*** 100
global cycling and, 94–96, ***95,*** 173
oceanic levels of, 85
photosynthesis and, 96, 107, 119
in plant decomposition, 29, 36
as poison and purifier, 199, 202, 208, 215
in radiometric dating, 79

Pacific Ocean, 50
Pacific plate, 50
Paleolithic Age, 69
Panisset, Maurice, 172–73
Pannekoek, Anton, 221
paramecia, 174
passive continental margin, 74–75
peat, ***25,*** 26, ***29***
food recycling by, 31
formation of, 28–30
as fuel, 37
mining of, 37–39, ***38***
types of, 32, ***33***
water retention by, 30, 36
Permian period, 61, 98, 100, 104
Perry, Ed, 91–93, ***92,*** 97–102, ***99,*** 214–15
petroleum, 93, 132, 149
Phanerozoic period, 94
phosphate, 40, 119, 154, 155, 162, 169
phosphorus, 188, 208
photosynthesis, 162, 163, 169, 170, 172, 202

calcium carbonate produced in, 164–65
carbon dioxide in, 96, 101, 155, 194–95, 200–201
chalk mud production and, 155–56
chemical reactions in, 194–95, 200–201
chlorophyll in, 118
in desktop biosphere, 186–87, ***189,*** 190, 194–95, 197
organic carbon produced in, 107, 155, 157
oxygen produced in, 96, 107, 119, 200–202
water in, 200–201
plankton, 78–79, 137, 164
plants:
decomposition of, 28–29, 36
as eukaryotes, 174
mineral conversion by, 130–31
transpiration by, 129–30
vascular, 130
of Vlieland's sand dunes, 126
plate tectonics, 48–54, ***49,*** 65, 89, 159, 163, 217
convection currents as driving force of, 51, 198
global-scale view of, 111–13
impact of, 50–51, 72
principle of, 49–50, ***49***
rock cycling and, 134, 135
see also subduction zones
platforms, carbonate, 158–62, 163
poisons, 189, 196–97, 208
polders, ***25,*** 26, 37, 39
polysaccharides, 139–40, ***142–43,*** 150, 155, 165
potassium, 188
Precambrian period, 98, 175
prokaryotic cells, 170, 173
proteins, 167, 168, 191, 192
protists, 172, 188, 190
pyrite, *see* sulphur, sulphide isotope of

quartz, 92–93

radiometric dating, 78–81, ***80,*** 217
rain forests, 129–30
Redfield, Arthur, 212
respiration, 96
Rhine river, ***27,*** 28, 114
rock cycling, 111–35, ***135,*** 162–63
description of, 112–13, 134–135, ***135***
life force and, 129–35
nutrients and, 127–29, 130, 135, 195–96
on Vlieland, 113
see also soil
rust, 196, 199

sahel, 131
Salter, J. W., 61
samphire, 126
Sandberg, Philip, 149, 150, 151, 152
sand dunes, 175
cycles of, 128–29
grasses of, 121–26, ***125***
microbial mats of, 117–21, 126, 129, 167
of Netherlands, 26, ***27,*** 28, ***29***
of Vlieland, 113–26
Scheldt river, ***27***
Schutter, Jaap, 24
science:
culture and, 224
disciplinary split in, 21–22, 42, 217–18
falsifiability in, 214, 216
integrating earth and life branches of, 217–18, 227
prophesies of doom by, 222
social, 224
as spark of hope, 223–24
theory and practice coupled in, 213–14
Scotland, 54–64, ***55, 56, 64,*** 88
Girvan district of, 54–64
subduction zone of, 62–64, 65, ***67***
tectonic motion of, 61–62
sea grass, 150–53
sediment, 70, 74, 78, 129, 134, 145
continental rise formed by, 75–76
life force and, 132–34, ***133***
ocean levels and, 84–85
plankton fossils in, 78–81, ***80***
subduction and, 52
sedimentary environment, 70, 71, 82, 148
sediment record, 77–78, 81, 226
seismology, 73–76, ***75,*** 81, 217
serpulid worms, 153
sewage, 40
sheep, 35

shrimp, 183–84, 186–90, 192
silicate minerals, 92, 133
silver, 188
slime, 165, 166–69
  as calcium carbonate precipitation inhibitor, 155, 167
  as glue for biosphere, 167
  human body functions of, 167–68, 169
  macromolecular, 131, 148, 150
  molecular properties of, 168
  in mud mounds, 175–79
  polysaccharides in, 155
  recycling of, 169
  water content of, 168
slugs, 168
social science, 224
sodium, 188
sodium carbonate, 136–37
soil, 175
  erosion of, 38, 85, 114, 129, 131–32, 134
  formation of, 129, 131, 196
  nutrients of, 127–29, 130–32
  transport of, 129, 131–32, 134
  weathering and, 130–31, 134
Sonea, Sorin, 172–73
South America, 49, 50
Southern Uplands, 62, 63, ***64***
Space Biospheres Ventures, 185
species diversity, 211
sphagnum moss, 28, 30–31
steam pump, 39
Stinchar Valley, 54–57, ***56, 64***
stromatolites, 163, 178
subduction zones, 50
  in Java, 51–54, 63, 64
  in Scotland, 62–64, 65, ***67***
sulphur, 188, 208
  carbon cycle coupled with, 102–6, ***104,*** 214, 216
  isotope reservoirs of, 93, 98–106, ***99, 104***
  sulphate isotope of, 93, 98–106, ***99, 104,*** 119, 197
  sulphide isotope of, 93, 98–106, ***99, 104,*** 119, 204
sulphuric acid, 146
symbiosis, 169–73

tectonics, *see* plate tectonics
Teilhard de Chardin, Pierre, 220
Texel, 114–15
transpiration, 129–30

uranium, 134, 189, 196
Utah, 131

Vail, Peter, 81–82, ***83,*** 84–85
Vail Curve, 84–85, ***86, 87***
van der Meulen, Frank, 114, 115
van Hinte, Jan, 73, 88
  deep-sea drilling expedition, 76–78
  on ocean levels and evolution, 86
  on seismology, 74
  Vail's hypothesis tested by, 81–82
vascular plants, 130
Venus, 167, 207
Vernadsky, Vladimir, 21, 212, 220–21
Viking Mars landing, 207
Vliehors, 116–17
Vlieland, ***27,*** 113–29, ***116, 125,*** 130, 131
  microbial mats of, 117–21, 126, 129, 167
  mud flats of, 126–27
  rock cycling on, 113
  sand dunes of, 113–26
  soil nutrients of, 127–29
volcanism, volcanoes, 48, 50, 51, 52, 53, 96, 145, 159, 162
Vos, Peter, 127

Wadden Zee mud flats, 126
water:
  drainage management systems for, 31, 34–37
  erosion by, 38, 85
  in photosynthesis, 194–95, 200–201
  in plant decomposition, 29
water cycle, 23, 129–30
water pollution, 40
weathering, 57–58, ***59,*** 129, 130–31
Whitfield, Michael, 94
Williams, Alwyn, 54, 60, 61
Wilson, Tuzo, 61
windmills, 36–37, 39
Wise, Sherwood, 76
worms, serpulid, 153

Young, J. Z., 212

zinc, 188